CONTINUATION METHODS

ACADEMIC PRESS RAPID MANUSCRIPT REPRODUCTION

Proceedings of a Symposium at
the University of Linz, Austria,
October 3-4, 1977.

CONTINUATION METHODS

EDITED BY

Hansjörg Wacker

Universität Linz
Institut für Mathematik
Linz, Austria

ACADEMIC PRESS New York San Francisco London 1978
A Subsidiary of Harcourt Brace Jovanovich, Publishers

ACADEMIC PRESS, INC.
111 Fifth Avenue, New York, New York 10003

United Kingdom Edition published by
ACADEMIC PRESS, INC. (LONDON) LTD.
24/28 Oval Road, London NW1 7DX

Library of Congress Cataloging in Publication Data

Symposium on Continuation Methods, Johannes Kepler Universität Linz, 1977.
Continuation methods.

1. Mathematics—Congresses. I. Wacker, Hansjörg, Date II. Title. III. Title: Continuation methods.
QA1.S899 1977 510 78-15678
ISBN 0-12-729250-0

PRINTED IN THE UNITED STATES OF AMERICA

CONTENTS

LIST OF CONTRIBUTORS

Numbers in parentheses indicate the pages on which authors' contributions begin.

J. C. ALEXANDER (37), Department of Mathematics, University of Maryland, College Park, Maryland 20742

F. J. DREXLER (69), Omptedastrasse 8, D-8 Munich 60, Federal Republic of Germany

J. HACKL (95), Wiener Allianz Lebensversicherungs-AG., Hietzinger Kai 101-105, Postfach 1500, A-1131 Vienna, Austria

G. HEINDL (129), Mathematisches Institut, Technische Universität, Barerstrasse 23, D-8 Munich 2, Federal Republic of Germany

H. JEGGLE (157), Fachbereich Mathematik, Technische Universität, Strasse des 17. Juni 135, D-1 Berlin 12, Federal Republic of Germany

M. PRÜFER (187), Institut für Angewandte Mathematik, Universität, Wegelerstrasse 6, D-53 Bonn, Federal Republic of Germany

J. W. SCHMIDT (215), Sektion Mathematik, Technische Universität, Mommsenstrasse 13, DDR-8027 Dresden, German Democratic Republic

HANSJÖRG WACKER (1,249), Institut für Mathematik, Universität Linz, A-4045 Linz, Austria

W. L. WENDLAND (277), Fachbereich Mathematik, Technische Hochschule, Kantplatz 1, D-61 Darmstadt, Federal Republic of Germany

E. ZARZER (249), Institut für Mathematik, Universität Linz, A-4045 Linz, Austria

W. ZULEHNER (249), Institut für Mathematik, Universität Linz, A-4045 Linz, Austria

PREFACE

This volume contains the Proceedings of the Symposium on Continuation Methods, held at the Johannes Kepler University in Linz, Upper Austria, October 3-4, 1977. The aim of the symposium was to bring together people who contribute to the development of continuation methods in various fields. Thus the participants of the symposium represent mathematical disciplines from topology to numerical analysis.

The symposium was opened with words of welcome by the Associate Dean, Professor Doctor Arno Schulz. I take this opportunity to thank the Deutsche Forschungsgemeinschaft (Federal Republic of Germany) and the Fonds zur Förderung der wissenschaftlichen Forschung (Austria) for their continuous support and the Austrian Federal Ministry for Science and Research (BMfWF) for financial support of the symposium.

The program committee consisted of the editor of this volume and Heinz W. Engl. All contributions in this volume have been refereed. I thank the referees J. B. Cooper, H. W. Engl, M. Feilmeier, C. Fenske, G. Heindl, W. Krabs, M. Schneider, H. Schwetlick, Hj. Stetter, and W. Zulehner.

I also wish to record my special thanks to the organizing committee, consisting of Ms. M. Traunmüller and Ms. W. Eidljörg, who also typed the manuscript. Many thanks also to Academic Press for publishing these proceedings and to Professor W. F. Ames for assisting with the publication arrangements.

A SUMMARY OF THE DEVELOPMENTS ON IMBEDDING METHODS

Hansjörg Wacker

Universität Linz
Institut für Mathematik
Linz, Austria

Nowadays an attempt to give a complete survey for even as small an aera of research like "Imbedding Methods" is almost impossible. Nevertheless it seems to be a useful task, like giving a starting point for an iteration process that could converge eventually to completeness. For readers not too familiar with our subject we shall first describe some points of interest - of course a task that cannot be done totally impartially - which might help to clarify the contributions made to this field during the last decades. To begin with an omission we shall never touch "Invariant Imbedding", a subject very close to ours (for a good presentation see for instance: G.H. Meyer, Initial Value Methods for Boundary Value Problems-Theory and Application of Invariant Embedding, Academic Press, 1973).

ISBN 0-12-729250-0

A. SOME ASPECTS INTERESTING IN CONNECTION WITH IMBEDDING METHODS

1. The Idea of the Imbedding Method. We want to solve the nonlinear equation $T_1(x)=o$, say in a Banach-space. Each local iteration method for the solution of this equation based on the Contraction Principle requires assumptions of the following kind (let $\bar{x}$ be the solution of the equation):

V1) Existence of a known starting point x_o sufficiently near to $\bar{x}$.

V2) Continuity of T_1, T_1', boundedness of T_1', validity of a Lipschitz-condition for T_1' (or continuity of T_1''), existence of a local inverse of T_1' at $x=x_o$.

V3) The assumptions V2) be true in a sufficiently large ball around the starting point x_o.

Then by the Contraction Principle one gets a constructive method with error bounds for obtaining the solution $\bar{x}$ which is at least locally unique. The imbedding method (continuation method, homotopy method) gives a possibility to dispense with the rather restrictive assumption V1).

To this end one imbeds the problem $T_1(x)=o$ into a famlily of problem $T(x,s)=o$, $s \in I:=[o,1]$, with $T(x,1) \equiv T_1(x)$ such that the problem $T(x,o)=o$ is easily solvable or even possesses a known solution. The basic idea consists in continuously transforming the solution $x(o)=x_o$ into the wanted solution $x(1)=\bar{x}$.

Examples for realizations of imbeddings

1) $T(x,s) \equiv T_1(x)+(1-s)T_1(x_o)$ (x_o is a reasonable initial guess)

2) Integral-equations:

$$x(t) = \int_o^s k(t,\tau,x(\tau))dt$$ (starting solution x_o=o)

3) Elliptic partial differential equations:

$$\Delta u = sf(x,y,u) \qquad (x,y) \in B \subset \mathbb{R}^2$$

$$u = \phi(x,y) \qquad (x,y) \in \partial B$$

(for s=o one has to solve a linear boundary value problem)

Problem 1: How to choose an imbedding which is adapted to a special equation with regard to computational effort, stability, avoidance of singularities.

2. *Numerical Methods.* The realization of an imbedding - that is the choice of a concrete homotopy - does not yield a numerical method to solve the family of equations. Without regard to the method of "Invariant Imbedding" not to be mentioned here at all, there are two main lines which now however can be discussed under a common aspect.

a. *Discrete methods.* After discretization of the parameter interval I (sometimes $[o,\infty)$) one gets a finite chain of problems:

$$T(x,s_n)=o, \quad o=s_o < s_1 < \ldots < s_N=1$$

Starting with the solution $x(o)=x_o$ of the first equation $T(x,o)=o$ the transition $x(s_n) \rightarrow x(s_{n+1})$ can be done by a special local iteration method. The whole process (if it works) is called a globalization of this local method.

As an example we present the globalization of the Newton method.

Theorem: Let the operator $T(x,s)$ be twice continuously differentiable in the sense of Frechet and in $K(x_o,R) \times I := \{ x \varepsilon X \mid \|x-x_o\| \leq R\} \times [o,1]$ the following assumptions be valid:

1) $T(x_o,o)=o$

2) $\bigvee_{A,B,M \geq o} \quad \bigwedge_{(x,s) \varepsilon K \times I}$ $\|T'_s(x,s)\| \leq A$, $\|T''_x(x,s)\| \leq B$, $\|T'_x(x,s)^{-1}\| \leq M$

3) for $\Delta s < S_o := \frac{2}{M^2BA}$, $R > AM(1 + \Delta s \sum_{k=o}^{\infty} (\frac{\Delta s}{S_o})^{2^k-1})$

Then:

for all equidistant stepsizes $\Delta s := s_{n+1} - s_n < S_o$ the local transition $x(s_n) \rightarrow x(s_{n+1})$ can be performed by the Newton method for all $n=o(1)(N-1)$ with the starting solution $x^o_{n+1} := x(s_n)$ at each level n.

The main idea of the step-method (discrete method) consists in forcing the Lipschitz-constant of the local iteration method below 1 by choosing Δs sufficiently small.

The theorem formulated above has two disadvantages. First it works with the exact solutions of the auxiliary problems $T(x,s_n)=o$ which can hardly be obtained in actual computation. Second the global estimates for $\|T'_s\|$, $\|T''_x\|$, $\|T'^{-1}_x\|$ are difficult to obtain and moreover not really needed in all of $K(x_o,R) \times I$ but only in a certain neighbourhood of the so-called homotopy path (s,x(s)).

Problem 2: State theorems which use only local estimates for $\|T'_s\|$, $\|T''_x\|$, $\|T'^{-1}_x\|$. Which local iteration methods should be globalized with regard to the compuational effort ?

In literature in almost all cases the stepsize $\Delta s_n := s_{n+1} - s_n$ is chosen in such a way that $x(s_n)$, or an approximation for it, lies in the region of attraction of the solution $x(s_{n+1})$ of $T(x,s_{n+1})=o$. It seems to be reasonable to discuss the computational effort in dependence of the stepsize. For a long period of time the general opinion was that imbedding methods should be only used in absence of a feasible starting value. It can be shown, however, that imbedding methods can also be used to accelerate the convergence of local iteration methods.

Problem 3: Construction of an applicable stepsize control which minimizes the computational effort of the whole process.

b. Davidenko's differential equation. By differention with respect to the parameter s one obtains the so-called Davidenko differential equation (D-DE):

$$T'_x(x(s),s)\frac{dx(s)}{ds} + T'_s(x(s),s) = \theta$$

Together with $x(o)=x_o$ $(T(x_o,o)=O)$ this is an initial value problem in a Banach-space. Both theoretically and numerically the implicit character of this equation turns out to be troublesome. By numerical integration of (D-DE) one obtains an approximation of the homotopy path .

Problem 4: Stepsize control for the numerical integration of (D-DE) to minimize the computational effort. Effective methods to cope with the implicit character of the equation.

Both stepsize methods (discrete methods) and methods for (D-DE) (continuous methods) can be summarized under the common aspect of predictor-corrector methods. $T(x,s)$ assumed to be sufficiently smooth one gets practicable methods if $T'_x(x(s),s)$ is regular respectively not ill-conditioned .

3. The Singular Situation ($T'_x(x,s)$ Not Regular).

a. Treatment as a bifurcation problem. $T: X \times I \to Y$, X, Y Banach-spaces, T sufficiently smooth and $T'_x(\bar{x},\bar{s})$, ($(\bar{x},\bar{s})$ be a point of the homotopy path), be a Fredholm operator of Noether-index o. Setting $x(\bar{s}+\Delta s) := \bar{x}+u$ one obtains the following equation for the continuation of the homotopy path:

$$\text{(C)} \quad Du := T'_x(\bar{x},\bar{s})u = [T'_x(\bar{x},\bar{s})u - T(\bar{x}+u,\bar{s}+\Delta s)] =: f(u,\Delta s)$$

One tries to find continuous projectors $P: X \to P(X)$ and $Q: Y \to Q(Y)$ with the same finite dimension for $P(X)$ and $Q(Y)$. Putting $u_1 := Pu$, $u_2 := (I-P)u$ there results out of (C):

$$\text{a)} \quad QDu_1 + QDu_2 = Qf(u_1+u_2,\Delta s)$$

$$\text{b)} \quad (I-Q)u_1 + (I-Q)Du_2 = (I-Q)f(u_1+u_2,\Delta s)$$

For a non-degenerate matrix representation of D, (I-Q) is

invertible on (I-P)X, one obtains uniquely out of b) a solution $u_2=u_2(u_1,\Delta s)$. This result substituted in a) yields for u_1 the socalled Bifurcation equation which in our case is finite dimensional. For each solution of this equation one obtains a possible continuation: $u=u_1(\Delta s)+u_2(\Delta s)$. Being interested in all solutions one has to follow not only the continuations branching to the right but also those branching to the left (see *c. Turning points*).

Problem 5: Numerical computation of a singular point $(\bar{x}=x(\bar{s}),\bar{s})$ with sufficient accuracy.

Examples for matrix representations of D

$$\begin{pmatrix} QDP & QD(I-P) \\ (I-Q)DP & (I-Q)D(I-P) \end{pmatrix}$$

for the special case $T:=I-A:X \to X$ and A compact with Rieszindex ν:

Example 1: Classical treatment of the bifurcation problem by employment of the simple null space.

$P_N:X \to N(D)$, $P_R:X \to R(D)$ yields with the setting $P:=P_N$, $Q:=(I-P_R)$ the following non-degenerate matrix respresentation:

$$D := \begin{pmatrix} O|_{P_N(X)} & O|_{(I-P_N)(X)} \\ O|_{P_N(X)} & P_RD(I-P_N)|_{(I-P_N)(X)} \end{pmatrix}$$

a) $(I-P_R)f(u_1+u_2,\Delta s) = o$

b) $P_RD(I-P_N)u_2 = P_Rf(u_1+u_2,\Delta s)$

Example 2: Employment of the generalized null space $N_\nu := N(D^\nu)$. $P_N : X \to N(D^\nu)$, $P_R \equiv I - P_N : X \to R(D^\nu)$ are continuous projectors by the theory of Riesz-Schauder. With $P := P_N$ and $Q := I-P$ one obtains the following non-degenerate representation:

$$D := \begin{pmatrix} P_N D P_N & P_N D(I-P_N) \\ (I-P_N) D P_N & (I-P_N) D (I-P_N) \end{pmatrix}$$

The commutability of the projectors with D yields:

a) $Du_1 = P_N f(u_1+u_2, \Delta s)$

b) $Du_2 = (I-P_N) f(u_1+u_2, \Delta s)$

Problem 6: Which matrix representation (induced by the projectors P,Q) should be chosen ? (Commutability of projectors and D, numerical feasability, stability, computational effort).

c. Change of parameters. By change of the parameter s to t with $s=\phi(x,t)$ (ϕ suitably chosen) one can suspend with the singularity of T'_x at a point $(\bar{x},\bar{s})$. One obtains:

$$\hat{T}(x,t) := T(x,\phi(x,t)) \equiv T(x,s)$$

$$\Longrightarrow \quad \hat{T}'_x(\bar{x},\bar{t}) = T'_x(\bar{x},\bar{s}) + T'_s(\bar{x},\bar{s}) \cdot \phi'_x(\bar{x},\bar{t})$$

Problem 7: Given a one dimensional imbedding $(s \in I \subseteq \mathbb{R}^1)$, how to choose $\phi(x,t)$ (also under numerical aspects) in the case of $\dim N(T'_x) > 1$?

d. Parameterization in the neighbourhood of a singular point $(\bar{x},\bar{s})$. We add a second equation $V(x,s,t)=o$ to $T(x,s)=o$ obtaining $P(z,t) \equiv (T(x,s),V(x,s,t) = o$. $z=z(t)=(x(t),s(t))$ satisfies $P(z(t),t)=o$. By choosing V suitably one gets a regular P'_x though T'_x is singular. To give an impression we quote the following results:

Theorem: $T:X \times I^k \to X$, X a Banach space, T sufficiently smooth, $T'_x(\bar{x},\bar{s})$ singular.

a) Let $\dim N(T'_x) = \operatorname{codim} R(T'_x) = k$. Then P'_x is non-singular iff

1) $\dim R(T'_s) = \dim R(V'_x) = k$

2) $R(T'_x) \cap R(T'_s) = \{o\} \wedge N(T'_x) \cap N(V'_x) = \{o\}$

b) In case of $\dim N(T'_x) > k := \operatorname{codim} R(T'_x)$, P'_x is singular.

For the important special case $\nu=1$ together with $\dim N(T'_x) = \operatorname{codim} R(T'_x) = 1$, one has to demand:

$$T'_s \notin R(T'_x) \quad \wedge \quad V'_x \notin R(T'^*_x)$$

On the basis of this theory one for instance gets a lot of numerical methods for the computation of turning points.

Problem 8: Choice of the operator V both to remove the singular situation and to obtain good numerical methods. How to handle the singular situation even in the case $k > 1$ with an imbedding parameter $s \in I \subseteq \mathbb{R}^1$ (and not $s \in I^k \subseteq \mathbb{R}^k$).

B. A HISTORICAL SURVEY ABOUT IMBEDDING METHODS

The basic idea of imbedding arose from the very obvious and widely used method to replace a large jump by a number of smaller ones. Therefore in my opinion it does not make sense to follow the idea of imbedding until the first origin. For instance there are connections with Analytic Continuation and more general with Perturbation Theory.

Two interesting aspects were recognized already in the Thirties. First the socalled Schauder's proof technique: Let $T(x,o)=o$ be a family of problems depending on the parameter $s \in I := [o,1]$. $S := \{s \in I / T(x,s)=o \text{ solvable}\}$. Then the following is true: $S \neq \emptyset \wedge S$ open $\wedge$ S closed $\Longrightarrow S=I$. Because the connected topological space I does not allow any nontrivial partitions in open-closed subspaces. This simple fact suggests to connect homotopically a problem to be solved with a problem already solved. Schauder [87] proved with this method the solvability of certain linear elliptic boundary value problems. At the same time Lahaye [58] got a connection to numerical analysis by a constructive contribution to the Implicit Function Theorem. Both papers had a number of predecessors; see for this the bibliography in the 1951 paper of Ficken [34] .

Friedrichs [36] took up these ideas in 195o in his "Lectures on Functional Analysis" and for the first time proposed a globalization for the Newton method. Under certain assumptions - $T'_x(x,s)$ regular etc. - it was possible to prove the existence of a positive lower bound for Δs_n therefore getting N finite in the chain of problems $T(x,s_n)=o$, $o=s_o< \ldots < s_N=1$. Further Friedrichs discussed the bifurcation problem (a necessary condition for it is: $T'_x(x,s)$ singular) and proposed an algorithm for the computation of the possible branches in $\mathbb{R}^n$. Ficken [34] 1951 continued the ideas of Friedrichs and got à priori bounds for the necessary number N of steps (more precisely, upper bound for N). With this at the same time he obtained a uniform stepsize Δs to solve the chain of problems $T(x,s_n)=o$, $n=o(1)N$ theoretically.

A new line of development was opened 1953 by Davidenko [17] , [18] with the proposal to differentiate the imbedded equation $T(x,s)=o$ with respect to the parameter s. Together with $x(o)=x_o$ this gives an initial value problem thus allowing the application of differential equation methods in abstract spaces. We already hinted the handicap of the Davidenko Differential Equation (D-DE) $T'_x\, x + T'_s = o$ to be implicit. Davidenko [19] 196o gave an application (inversion of matrices).
Schidlovskaja [9o] 1958 continued his ideas. In 1958 Gavurin [37] was able to work with stability methods by the transition from $[o,1]$ to $[o,\infty)$. Yakovlev [11o] 1963 worked on the theoretical solvability of (D-DE), Kizner [53] 1964, Bittner [7] 1965 and in particular Kleinmichel [54] 1967 already obtained numerical algorithms (and results) by integrating (D-DE) with integration methods like Runge-Kutta etc.

Independently from these two both theoretically and numerically rather advanced lines of research, some more heuristically written papers which nevertheless gave a strong push to the idea of imbedding were published: above all the frequently quoted paper by Freudenstein-Roth [35] 1963. The algebraic systems solved there are still used nowadays as a kind of touchstone for the presentation of new imbedding algorithms. Next to this paper one should mention that of Deist-Sefor [21] 1967.

The singular situation was attacked again. In 1966 Anselone-Moore [5] proposed to suspend with the singularity of $T'_x(\bar{x},\bar{s})$ by a change of the parameter: $s=\phi(t,x)$, though under the somehow restrictive assumption $\dim N(T'_x(\bar{x},\bar{s}))=1$. Davis [20] 1966 gave a generalization for every finite dimension p but with the assumption $s \in \mathbb{R}^p$ which brings in some new problems: without difficulty the parameter s of imbedding can be made multidimensional at the beginning of the process but in general nothing can be said about $\dim(T'_x)$ à priori.

These general developments were summarized in a certain way by the paper of Meyer [66] 1968. Both the solution of (D-DE) and the globalized Newton method were discussed theoretically and numerically. A connection between these two lines was constructed. For the solution of the singular situation Meyer hints the possibility of a change of parameter though he himself discusses a kind of perturbed Newton-iteration using a certain partial ordering. The stepsize Δs

which can be regarded here as the parameter h of discretization for a numerical integration method of order $O(h^p)$, can be (theoretically) computed with the aid of certain global constants like bounds for $\|T''_x\|$, $\|T'_s\|$, $\|T'^{-1}_x\|$.

During these years a number of papers was written about special applications of the imbedding method. Roberts-Shipman [83] 1967, [85] 1968, Roberts-Shipman-Roth [84] 1968 were working on shooting methods for two point boundary value problems (more precisely: the authors combined shooting methods and the continuation method thus obtaining the socalled multiple shooting method).
Bosarge [1o] 197o, [11] 1971, Zidkov-Puzynin [111] 1968, and Wacker [1o1] 197o were solving two point boundary value problems, Bosarge and Zidkov-Puzynin with continuous methods (D-DE), Wacker with discrete methods (step methods). For a class of integral- and integrodifferential equations of Fredholm type Wacker[99] 1969, [1oo] 197o, [1o2] 1972 proposed imbedding methods adapted to the structure of these problems. Broyden [13] 1969 presents the idea of a predictor. As a starting value for the next step at level n+1 he does not use the numerical solution of $T(x,s_n)=o$ but he computes by quadratic interpolation (more precisely, extrapolation) an approximation of higher order. Furthermore he proposes a kind of step size control by choosing the stepsize Δs_n in dependence of the curvature of the homotopy path respectively of its approximation. To reduce the computational effort the inverse of $T'_x(x_n,s_n)$ is calculated recursively out of its

predecessors. The numerical solution of transcendental systems was discussed under the aspect of imbedding methods. Laasonen [57] 197o pointed out the importance of complex valued computation in step methods in the neighbourhood of singular points. Baranov [6] 1972 integrated (D-DE) to get complex roots. Petry [72] 197o, [73] 1971 combines imbedding with monotonicity. Branin [12] 1972 takes up the ideas of Broyden concentrating on the numerical treatment of singular points. For the special homotopy $T(x,s) \equiv x-sT(x)$ the (D-DE) is integrated numerically adapting the orientation of the homotopy path between two zeroes of det T'_x to obtain stability of the integration process. This works as a predictor-corrector-process.

In the book of Ortega-Rheinboldt [71] 197o for the first time a survey over some important results in the field of imbedding methods was given. For instance Avila, Bosarge, Ficken, Kleinmichel, Meyer etc. See pp. 132-14o, 23o-236, 334-341.

Avila [1] 1971, [2] 1974, Boggs [8] 1971 , [9] 1973, Feilmeier [31] 1971, [3] 1972, Wacker [1o4] 1972, [I] 1973 were discussing the more general aspects of the method itself: Avila and Wacker globalized (discretely) the Newton method respectively the modified Newton method, Avila taking into account for the first time the apsect of implementability (in actual computation it is impossible to work with points lying exactly on the homotopy path). Wacker used bifurcation theory to treat the singular situation. For the matrix representation of the Fredholm operator $T'_x(\bar{x},\bar{s})$ he proposed to work with the generalized null space and range instead of the simple null space and range - a device suggested by some results about numerical approximation (see for example Jeggle [47] 1972, and more recently [48] 1977). Boggs used A-stable integration techniques for the solution of (D-DE). Replacing the computation of $T'_x(\bar{x},\bar{s})$, respective of its inverse, by a consistent approximation he got a kind of Quasi-Newton method. Feilmeier discussed connections between imbedding methods and projection methods. Feilmeier-Wacker [32] 1972 presented a special predictor-corrector method adapted to hybrid computation.

In his survey about some developments of imbedding Wasserstrom [1o8] 1973 puts his main interest in the comparison between imbedding methods and iteration methods thus getting for instance predictor-corrector methods. Because of its presentation - the discussion is led by means of some important classes of problems, for instance two point boundary value problems, eigenvalue problems etc., the paper should be interesting for engineers too.

In the following years imbedding has been applied in various fields of problems. Least squares problems: Kirzenblat-Chetrit [52] 1974, optimization problems: Glashoff [38] 1973. A series of papers is devoted to Multiple Shooting with concrete applications in praxis: Deuflhard [23] 1974, Deuflhard-Pesch-Rentrop [24] 1976, Oberle [69] 1976, [7o] 1977, Dickhoff-Lory-Oberle-Pesch-Rentrop [22] 1977,

Lory [61] 1977, Seydel [95] 1977. Scheuring [88] 1974 solved nonlinear elliptic systems with discontinuous coefficients (a generalization of the Hilbert problem) numerically by imbedding. Hachtel-Mack [39] 1974 integrated (D-DE) with methods for stiff differential equations. Edelen [28] 1976 handled (D-DE) by Lyapunov's methods. Drexler [26] 1976 succeeded in adapting imbedding to the structure of polynomial systems, getting a method that proved to be superior for instance to deflation methods. Wacker [105] 1974, [106] 1977 worked on the globalization of some classes of local iteration methods and the possibility of alternating between local methods during the process of imbedding. Schwetlick [93] 1975, [94] 1976 proposed - besides other results - a very clear formulation of the connection between discrete and continuous imbedding methods by using the concept of a predictor-corrector scheme. For instance one starts with the numerical integration of (D-DE) from s_n to s_{n+1} by Euler's method (predictor), taking then the Newton method (corrector). In the sequel Menzel-Schwetlick [63] 1976 carried over the well-known definitions of order of an integration method to imbedding methods. Schmidt [92] 1975 pointed out the aspect of "Finitization" of an imbedding process. Hackl [41] 1977 combined optimization and imbedding under the aspect of the damped Newton method. Heindl [46] 1977 presented a descent algorithm for the product space problem $\|H(x,s)\|_2^2 = \min$! with $H(x,s):=(T(x,s),m(1-s))^T$ (see also Leder). Eigenvalue problems: Hayes-Wasserstrom [44] 1976.

The main interest of research during the last years in my opinion lies in the following three areas: stepsize control, numerical treatment of the singular situation, constructive topological methods.
In the beginning almost all papers which gave stepsize estimations constructed an upper bound S_o which had been got theoretically by the aid of global constants such that all transitions $x(s_n) \to x(s_{n+1})$ in the chain $T(x,s_n)=o$, $n=o(1)(N-1)$ were feasible for any stepsize $o < \Delta s_n < S_o$. See for instance [34] , [66] , [1] , [1o4] . The aim was to bring $x(s_n)$ - the approximation for $x(s_n)$ - into the region of attraction of the solution $x(s_{n+1})$ of the next equation $T(x,s_{n+1})=o$. This intention remained valid even in some recent papers, for example Rheinboldt [77] 1975, where upper and lower bounds for the local radius of convergence for the Newton method were computed or Hejer [45] 1976, who defined in a general way the radius of convergence for an iteration process with special application to (D-DE).
The philosophy common to all these papers was the assumption that a maximal (feasible) stepsize means minimal computational effort. But there is neglected that the number of iterations at each level n increases as the quality of the respective starting value drops (for instance $\tilde{x}(s_{n-1})$) and is therefore an increasing function of the stepsize Δs_n.
Ribarić-Seliškar [81] 1974 were the first ones to investigate the connection between stepsize and computational effort. Leder [6o] 1974 and more recently Heindl [46] 1977 transformed the equation $T(x,s)=o$ into a minimization problem $\|H(x,s)\|_2^2=\min!$ with $H(x,s):=(T(x,s),m(1-s))^T$, $m > o$. By this

idea the connection was found to self steering descent methods like Goldstine-Armijo. On the basis of global estimations Wacker [1o6] 1976 presented an effort-minimizing model for globalizations of minimal residue type. Though confessing to the philosophy of maximal step size Deuflhard [25] 1976 presented a predictor-corrector stepsize strategy that is steered by local estimates and therefore in actual computation proves to be superior to global strategies. Wacker-Zarzer-Zulehner [1o7] 1977 showed both theoretically and by tests for the Newton method and the modified Newton method that the computational effort is reached at an optimal stepsize $\overline{\Delta s}$ lying clearly below S_o. These results are backed by a paper of Kung [56] 1976, where also aspects of complexity are discussed. Hackl-Wacker-Zulehner [42] 1978 generalized a more refined implementable version of this model for any one-stage iteration process. Out of this results the possibility to accelerate the convergence of an iteration process by imbedding. It remains the question for all these models how precise they approximate reality. Besides it is not possible to decide only by functional-analytic-numerical methods (number of iterations) because there are aspects of complexity to be discussed (number of function evaluations).

The numerical treatment of the singular situation is connected so strongly with bifurcation theory that a clear seperation does not seem possible. Here we try to confine ourselves to contributions having in mind aspects of imbedding. First of all one has to make sure that one has reached a singular point. Seydel [95] 1977 presented an algorithm for the numerical computation of bifurcation points especially for two point boundary value problems. If one is content to get one solution at all at s=1 the idea "to jump over" singular points is near at hand. This can be performed by multi-parameter imbedding (we worked successfully with this method) or by enforcing uniqueness. For instance one can work only in the range of $T'_x(\bar{x},\bar{s})$ - $(\bar{x},\bar{s})$ be the singular point - with the additional requirement that $\|u\|$ be minimal (by use of the Moore-Penrose-pseudo-inverse that part of the continuation of x at $(\bar{x},\bar{s})$, u, is skipped which does not lie in $R(T'_x)$). See for instance Deuflhard [23] 1974, Deuflhard [25] 1977. Engl-Wacker-Zarzer [3o] 1977 and Hackl-Wacker-Zarzer-Zulehner [4o] 1977 discussed some aspects of bifurcation like how to choose a matrix representation for $T'_x(\bar{x},\bar{s})$ or the numerical solution of the nonlinear bifurcation equation by a kind of re-imbedding.

The idea of a change of parameter by Anselone-Moore was extended by Engl [29] 1976 to each finite dimension of $N(T'_x)$. The dimension of the imbedding parameter s resp. t (the new parameter) remains one. In comparison with this the suggestions of Davis [2o] 1966 and more recently Keller [49] 1977 have the disadvantage that the respective dimension k of $N(T'_x)$ has to be managed by a corresponding dimension of the imbedding parameter s: $s \in \mathbb{R}^k$.

On the background of the papers by Broyden [13] 1969, Thurston [97] 1969, Simpson [96] 1975, Kubiček [55] 1976, Chua-Ushida [15] 1976 algorithms were presented for weak singularities (for instance, turning points):Menzel-Schwetlick [62] 1975, Pönisch-Schwetlick [74] 1977, Menzel [64], [65]

1978. These algorithms make it possible to compute either a turning point itself or a good approximation of the homotopy path, or - by a jump - a solution of T(x,1)=o.

Topological methods were discussed by Scarf [86] 1967 who presented a numerical method for the computation of fixed points using the Brower Fixed Point Theorem (more precisely, Sperner's Lemma). The algorithm works well for low dimensions but the computational effort increases strongly with the dimension. An advantage of this kind of method lies in the independence from differentiability assumptions. A series of papers has been published until now, see for example Todd [98] 1976 . Kellogg-Li-Yorke [5o] 1976, [51] 1977 presented algorithms using the No-Retraction-Theorem; for instance, using the special homotopy T(x,s) ≡ s T(x)-x and integrating by Euler's method one gets the Newton method. Their methods still work at higher dimensions (say n=6o). Chow-Mallet Paret-Yorke [14] 1977 connected the problem T(x,1)=o not only with one problem T(x,o)=o but with a whole family T(x,o,a)=o, a ∈ A. Choosing randomly a value a ∈ A they get a smooth homotopy path that leads with "probability 1" to a fixed point. A parameterized version of Sard's Theorem and of course degree theory are the theoretical tools they used. Applications were given to two point boundary value problems. Alexander-Yorke [3] 1977, Alexander [4] 1977 put their main interest into the theoretical background of this subject, mainly: whether the zeroes of two functions are connected by suitable zero sets across the homotopy cylinder, whether the connecting set is smooth, so that it can in

principle be traced by constructive methods. Alexander gave some ideas how to get a working algorithm. Here is the hope that on this solid basis numerical methods can be found which possibly are superior to classical bifurcation methods.
Using suitably adapted methods of triangulation Prüfer [71] 1977 presented a numerical method by which it is possible to follow even unstable branches in the neighbourhood of branching points. Using classical methods (matrix representation) this is a tough thing to do. For an alternate method see the algorithm of Scheurle [89] 1977 where only stable branches are selected.

We started this survey with Schauder's proof technique applied to a class of linear elliptic differential equations. Crandal-Rabinowitz [16] 1975 discussed the possibility of continuation for positive solutions of nonlinear elliptic eigenvalue problems. Wendland [1o9] 1977 considered Riemann-Hilbert boundary value problems with semilinear equations in the Douglis normal form and linear boundary and side conditions. This class of problems is discretized by a Finite Element Method and linearized by a globalized Newton method with uniform convergence for all sufficiently small mesh sizes. The linear problems are solved by a Ritz-Galerkin method which produces sparse matrices.
The future of imbedding methods can be seen in these three fields discussed above: stepsize control, numerical treatment of the singular situation, numerical methods on the base of topology. Further possibilities lie in constructing problem-adapted homotopies for special classes.

REFERENCES

[1] Avila, J. H., Continuation methods for nonlinear equations, Technical Report TR-142, University of Maryland, Computer Science Center (1971), 1-138.

[2] Avila, J. H., The feasibility of continuation methods for nonlinear equations, SIAM J. Numer. Anal. 11 (1974), 102-120.

[3] Alexander, J. C., Yorke, J. A., The homotopy continuation method: numerically implementable topological procedures, University of Maryland, Rep. MD-77-19-JCA (1977), 1-26

[4] Alexander, J. C., The topological theory of an Embedding Method, Symposium "Imbedding Methods for the solution of nonlinear problems", Univ. Linz, Math. Inst. (1977), 1-29.

[5] Anselone, P. M., Moore, R. H., An Extension of the Newton-Kantorovic Method for solving nonlinear equations with an application to elasticity, J. Math. Anal. and Appl. 13 (1966), 476-501.

[6] Baranov, A. V., A method of calculating the complex roots of a system of nonlinear equations, Zh. vychisl. Mat. mat. 12 (1972), 199-203.

[7] Bittner, L., Einige kontinuierliche Analogien von Iterationsverfahren, ISNM Vol. 7 (1965), 114-135.

[8] Boggs, P. T., The solution of nonlinear systems of equations by A-stable integration techniques, SIAM J. Numer. Anal. 8, Nr. 4, (1971), 767-785.

[9] Boggs, P. T., On the continuous analogs of quasi-Newton methods, Math. Dept. Rensselaer Polytechn. Inst. Troy, New York 12181 (1973), 1-14.

[10] Bosarge, W. E. jr., A continuous analog for some Newton's method convergence theorems, IBM Scientific Center Rep. Nr. 320.2381 (1970), 1-15.

[11] Bosarge, W. E. jr., Iterative continuation and the solution of nonlinear two-point boundary value problems, Num. Math. 17 (1971), 268-283.

[12] Branin, F. J. jr., Widely convergent method for finding multiple solutions of simultaneous nonlinear equations, IBM J. Res. Develop. 16 (1972), 504-522.

[13] Broyden, C. G., A new method of solving nonlinear simultaneous equations, Comp. J. 12 (1969), 94-99.

[14] Chow, S. N., Mallet-Paret, J., Yorke, J. A., Finding zeroes of maps: homotopy methods that are constructive with probability one, Univ. of Maryland, College Park, Proj. MCS 76-24432 (1977), 1-25.

[15] Chua, L. O., Ushida, A., A switching-parameter algorithm for finding multiple solutions of nonlinear resistive circuits, Circuit Theory and Appl. 4 (1976), 215-239.

[16] Crandall, M. G., Rabinowitz, P. H., Some continuation and variational methods for positive solutions of nonlinear elliptic eigenvalue problems, Arch. Rat. Mech. a. Anal. 58 (1975), 207-218.

[17] Davidenko, D. F., Über eine neue Methode zur numerischen Lösung von Systemen nichtlinearer Gleichungen, DAN 88, Nr. 4 (1953), 601-604.

[18] Davidenko, D. F., Über eine Näherungslösung von Systemen nichtlinearer Gleichungen, Ukr. math. Journal 5, Nr. 2 (1953), 196-206.

[19] Davidenko, D. F., Inversion of matrices by the method of variation of parameters, Soviet Math., Vol. 1 (2) (1960), 279-282.

[20] Davis, J., The solution of nonlinear operator equations with citical points, Dept. of Math., Oregon state Univ. Technical Report Nr. 25 (1966), 1-69.

[21] Deist, F. H., Sefor, L., Solution of systems of nonlinear equations by parameter variation, Computer J. 10 Nr. 1 (1967), 78-82.

[22] Diekhoff, H.-J., Lory, P., Oberle, H. J., Rentrop, P., Seydel, R., Comparing routines for the numerical solution of initial value problems of ordinary differential equations in Multiple Shooting, Num. Math. 27 (1977), 449-469.

[23] Deuflhard, P., A modified Newton method for the solution of ill-conditioned systems of nonlinear equations with application to Multiple Shooting, Num. Math. 22 (1974), 289-315.

[24] Deuflhard, P., Pesch. H.-J., Rentrop, P., A modified continuation method for the numerical solution of nonlinear two-point boundary value problems by shooting techniques, Num. Math. 26 (1976), 327-343.

[25] Deuflhard, P., A stepsize control for continuation methods with special application to Multiple Shooting techniques, Habil. TU München, Preprint TUM Nr. 7627, (1976), 1-76.

[26] Drexler, F. J., A Homotopy-Method for the calculation of all zeros of zero dimensional polynomial ideals, Symposium "Imbedding Methods for the solution of non-linear problems", Univ. Linz, Math. Inst. (1977), 1-25.

[27] Eaves, B., Saigal, R., Homotopies for computation of fixed points on unbounded regions, Math. Programming 3 (1972), 225-237.

[28] Edelen, D. G. B., 1) On the construction of Differential Systems for the Solution of nonlinear Algebraic and Transcendental Systems of Equations, 67-84,
2) Differential Procedures for Systems of Implicit Relations and Implicitly Coupled nonlinear Boundary Value Problems, 85-95,
in: Numerical Methods for Diff. Syst. (eds. Lapidus-Schiesser), Academic Press (1976).

[29] Engl, H., On the change of parameters in continuation methods, Preprint Univ. Linz, Math. Inst., Bericht Nr.58 (1976); abstract in SIAM Review 19 (1977), 767.

[30] Engl, H., Wacker, Hj., Zarzer, E., Bemerkungen zur Aufwandsminimierung bei Stetigkeitsmethoden sowie Alternativen bei der Behandlung der singulären Situation, ISNM 38 (1977), 175-193.

[31] Feilmeier, M., Eine Verallgemeinerung der Friedrichsschen Stetigkeitsmethode, ZAMM 51 (1971), T18 - T19.

[32] Feilmeier, M., Wacker, Hj., Zur numerischen Praxis von Einbettungsmethoden, ZAMM 52 (1972), T200-202.

[33] Feilmeier, M., Numerische Aspekte bei der Einbettung nichtlinearer Probleme, Computing 9 (1972), 355-364.

[34] Ficken, F. A., The continuation method for functional equations, Comm. Pure Appl. Math. 4 (1951), 435-456.

[35] Freudenstein, F., Roth, B., Numerical solution of systems of nonlinear equations, J. Assoz. Comp. Math. Vol. 10, Nr. 4 (1963), 550-556.

[36] Friedrichs, K. O., Lectures on Functional Analysis, (Lecture Notes), Inst. f. Math. and Mech., New York University, Chapter VII, X (1950).

[37] Gavurin, M. K., Nonlinear functional equations and Continuous analoga of iterative methods, Isv. Vyssh. Uchebn. Zaved. Matematika 6, Nr. 5 (1958), 18-31.

[38] Glashoff, K., Eine Einbettungsmethode zur Lösung nichtlinearer restringierter Optimierungsprobleme, Oper. Res. Verfahren XVIII (1973), 92-102.

[39] Hachtel, G. and Mack, M., A pseudo dynamic method for solving nonlinear-algebraic equations, in: Stiff Diff. Equat. (ed. R. A. Willoughby), Plenum Press, New York - London (1974), 135-150.

[40] Hackl, J., Wacker, Hj., Zarzer, E., Zulehner, W., Über die Aufspaltung eines singulären Operators beim Bifurkationsproblem, Preprint Nr. 72, Math. Inst. d. Univ. Linz (1977), 1-20.

[41] Hackl, J., Solution of optimization problems with nonlinear restrictions via continuation methods, Symposium "Imbedding Methods for the Solution of nonlinear problems", Univ. Linz, Math. Inst. (1977), 1-32.

[42] Hackl, J., Wacker, Hj., Zulehner, W., Implementierbare Schrittweitensteuerungen bei globalisierten Iterationsverfahren, Preprint Nr. 101, Univ. Linz, Math. Inst. (1978), 1-25.

[43] Haselgrove, C. B., A Solution of Nonlinear Equations and of Differential Equations with Two-Point Boundary Conditions, Comp. J. 4 (1961), 255-259.

[44] Hayes, L., Wasserstrom, E., Solution of non-linear eigenvalue problems by the continuation method, J. Inst. Math. Appl. 17 (1976), 5-14.

[45] Heijer, C., Iterative solution of nonlinear equations by imbedding methods (Diss.), stichting mathematisch centrum NW 32/76 (1976), 1-45.

[46] Heindl, G., Some remarks of the convergence of descent methods, Symposium "Imbedding Methods for the solution of nonlinear problems", Univ. Linz, Math. Inst. (1977), 1-28.

[47] Jeggle, Hj., Über die Approximation von linearen Gleichungen 2ter Art und Eigenwertprobleme in B-Räumen, Math. Z. 124 (1972), 319-342.

[48] Jeggle, Hj., Existence and discrete approximation of bifurcation points, Symposium "Imbedding Methods for the solution of nonlinear problems", Univ. Linz, Math. Inst. (1977), 1-30.

[49] Keller, H. B., Numerical solution of bifurcation and nonlinear eigenvalue problems, to appear in: Appl. of Bifurcation Theory, Academic Press, New York (1977), 1-26.

[50] Kellogg, R. B., Li, T. Y., Yorke, J., A constructive proof of the Brouwer fixed-point theorem and computational results, SIAM J. Num. Anal. Vol. 13, Nr. 4 (1976) 473-483.

[51] Kellogg, R. B., Li, T. Y., Yorke, J., A method of continuation for calculating a Brouwer fixed point, in: Fixed Point-Algorithms and Applications (ed. S. Karamadian), Academic Press, New York (1977).

[52] Kirszenblat, A., Chetrit, M., Least squares nonlinear parameter estimation by the iterative continuation method, technical notes AIAA Journal, Vol. 12, Nr. 12 (1974), 1751-52.

[53] Kizner, W., A numerical method for finding solutions of nonlinear equations, SIAM J. Appl. Math., Vol. 12, Nr. 2 (1964), 424-428.

[54] Kleinmichel, H., Stetige Analoga und Iterationsverfahren für nichtlineare Gleichungen in Banachräumen, Math. Nachr. 37 (1968), 313-344.

[55] Kubicek, M., Algorithm 502 - Dependence of Solution of Nonlinear Systems on a Parameter [C5], ACM Transact. on Math. Software, Vol. 2, Nr. 1 (1976), 98-107.

[56] Kung, H. T., The complexity of obtaining starting points for solving operator equations by Newton's method, in: Analytic Computational Complexity (ed. Traub), Mellon, Univ. Pittsburgh, Academic Press (1976), 35-55.

[57] Laasonen, P., An imbedding method of iteration with global convergence, Computing 5 (1970), 253-258.

[58] Lahaye, E., Une méthode de résolution d'une catégorie d'équations transcendentes, C. R. Acad. Sci. Paris, Nr. 198 (1934), 1840-42.

[59] Leder, D., Zur Lösung nichtlinearer Gleichungssysteme mittels diskreter Einbettungsmethoden, Diss. TU Dresden, Sekt. Math. (1974), 1-72.

[60] Leder, D., Automatische Schrittweitensteuerung bei global konvergenten Einbettungsmethoden, ZAMM 54 (1970), 319-324.

[61] Lory, P., Hinreichende Bedingungen für die Durchführbarkeit von Einbettungsmethoden mit Berücksichtigung der Mehrzielmethode, Symposium "Einbettungsmethoden bei nichtlinearen Problemen", Univ. Linz, Math. Inst., (Vortrag)(1977).

[62] Menzel, R., Schwetlick, H., Zur Behandlung von Singularitäten bei Einbettungsalgorithmen, Preprint, Math. Sekt., TU Dresden, I:1-15, II:1-15 (1975).

[63] Menzel, R., Schwetlick, H., Über einen Ordnungsbegriff bei Einbettungsalgorithmen zur Lösung nichtlinearer Gleichungen, Computing 16 (1976), 187-199.

[64] Menzel, R., Über die Lösung nichtlinearer Gleichungssysteme bei regulärer und schwach singulärer Einbettung, Diss. A., TU Dresden (1978).

[65] Menzel, R., Ein implementierbarer Algorithmus zur Lösung nichtlinearer Gleichungssysteme bei schwach singulärer Einbettung, to appear in: Beitr. Numer. Math., (1978).

[66] Meyer, G., On solving nonlinear equations with a one-parameter operator imbedding, SIAM J. Num. Anal. 5 (1968), 739-752.

[67] Miele, A., Naqvi, S., Levy, A. V., Modified quasilinearization method for solving nonlinear equations, Rice Univ., Aero-astronautics Report Nr. 78 (1970), 1-22.

[68] Miele, A., Naqvi, S., Levy, A. V., Iyer, R. R., Numerical solution of nonlinear equations and nonlinear, two-point boundary-value problems, Adv. Contr. Syst. 8 (1971), 189-225.

[69] Oberle, H.-J., Numerical computation of Minimum fuel space travel problems by multiple shooting, TUM Abt. Math. Ber. 7635 (1976), 1-25.

[70] Oberle, H.-J., Numerische Behandlung singulärer Steuerungen mit der Mehrzielmethode am Beispiel der Klimatisierung von Sonnenhäusern, Preprint TU München, Nr. 7734 (1977), 1-126.

[71] Ortega, J. M. and Rheinboldt, W. C., Iterative Solution of Nonlinear Equations in Several Variables, Academic Press, New York, London (1970).

[72] Petry, W., Eine Variante des Newtonschen Verfahrens, ZAMM 50 (1970), T74-T75.

[73] Petry, W., Eine Verallgemeinerung des Newtonschen Iterationsverfahrens, Computing 7 (1971), 25-45.

[74] Pönisch, G. and Schwetlick, H., Ein lokal überlinear konvergentes Verfahren zur Bestimmung von Rückkehrpunkten implizit definiter Raumkurven, Preprint, Math. Sekt. d. TU Dresden (1977), 1-16.

[75] Prüfer, M., Calculating global bifurcation, Symposium "Imbedding Methods for the solution of nonlinear problems", Univ. Linz, Math. Inst. (1977), 1-27.

[76] Reiss, E. L., Imperfact Bifurcation, Preprint: Courant Inst. of Math. Science, New York Univ. (1976/77), 1-29.

[77] Rheinboldt, W. C., An adaptive continuation process for solving systems of nonlinear equations, Univ. of Maryland, College Park, Technical Report TR-393, NSF (1975), 1-23.

[78] Rheinboldt, W. C., Numerical methods for a class of finite dimensional bifurcation problems, Univ. of Maryland, College Par, Technical Report TR-490 (1976), 1-21.

[79] Rheinboldt, W. C., Numerical continuation methods for finite element application, Univ. of Maryland, Technical Report TR-454 (1976), 1-28.

[80] Rheinboldt, W. C., On the solution of some nonlinear equations arising in the application of finite element methods, in: Math. of Finite Elements, (ed. J. Whiteman), Academic Press, London (1976), 465-482.

[81] Ribarić, M., Seliškar, M., On optimization of stepsize in the continuation method, Mathematica Balcanica 4.97 (1974), 517-521.

[82] Riks, E., The application of Newton's method to the problem of elastic stability, Trans. ASME, Serie E, J. Appl. Mech. 39 (1972), 1060-1065.

[83] Roberts, S. M., Shipman, J. S., Continuation in Shooting methods for two-points boundary value problems, J. of Math. Anal. Appl. 18 (1967), 45-58.

[84] Roberts, S. M., Shipman, J. S., Roth, C. V., Continuation in Quasilinearization, J. of Optimization Theory and Appl. Vol. 2, Nr. 3 (1968), 164-178.

[85] Roberts, S. M., Shipman, J. S., Justification for the continuation methods in two-point boundary value problems, J. of Math. Anal. and Appl. 21 (1968), 23-30.

[86] Scarf, H., The approximation of fixed points of a continuous mapping, SIAM J. Appl. Math. 15 (1967), 1328-43.

[87] Schauder, J., über lineare elliptische Differentialgleichungen 2ter Ordnung, Math. Z. 38 (1934), 257-282.

[88] Scheuring, H. T., Einbettung bei elliptischen Randwertproblemen mit unstetigen Koeffizienten und nichtlinearen Konjugiertheitsbedingungen, ZAMM 54 (1974), T233-234.

[89] Scheurle, J., Ein selektives Projektions-Iterationsverfahren und Anwendungen auf Verzweigungsprobleme, Num. Math. Fasc. 1 (1977), 11-35.

[90] Schidlovskaja, N. A., Anwendung der Methode der Differentiation nach einem Parameter zur Lösung nichtlinearer Gleichungen in Banachräumen, (russ.), Leningrad Gos. Univ. Uceu. Zap., Ser. Mat. Nauk. 33 (1958), 5-17.

[91] Smale, S., Convergent processes of price adjustment and global Newton methods, J. Math. Econ. 3 (1976), 1-14.

[92] Schmidt, J. W., Finitisierung und Globalisierung bei Iterationsverfahren zur Lösung nichtlinearer Gleichungssysteme, Bericht der Techn. Univ. Dresden, Sekt. Math. (1975), 1-21.

[93] Schmidt, J. W., Selected contributions to imbedding methods for finite dimensional problems, Symposium "Imbedding Methods for the solution of nonlinear problems", Univ. Linz, Math. Inst. (1977), 1-33.

[94] Schwetlick, H., Ein neues Prinzip zur Konstruktion implementierbarer, global konvergenter Einbettungsalgorithmen, Beitr. z. Num. Math. 4 (1975), 215-228.

[95] Schwetlick, H., Ein neues Prinzip der Konstruktion implementierbarer, global konvergenter Einbettungs-algorithmen (Testbeispiele), Beitr. z. Num. Math. 5 (1976), 201-206.

[96] Seydel, R., Numerische Berechnung von Verzweigungen bei gewöhnlichen Differentialgleichungen, Preprint, TU München, Nr. 7736 (1977). 1-83.

[97] Simpson, R. B., A method for the numerical determination of bifurcation states, SIAM J. Numer. Anal. 12 (1975), 439-451.

[98] Thurston, G. A., Continuation of Newton's method through bifurcation points, J. Appl. Mech. 36 (1969), 425-430.

[99] Todd, M. J., The computation of fixed points and application, Lecture Notes in Economics a. Math. Systems, Vol. 124, Springer, Berlin (1976).

[100] Wacker, Hj., Eine Methode zur numerischen Lösung von nichtlinearen Gleichungen Fredholmscher Art, ZAMM 49, Heft 11 (1969).

[101] Wacker, Hj., Ein Stufenverfahren zur Lösung nicht-linearer Integralgleichungen, ZAMM 50 (1970), T78-79.

[102] Wacker, Hj., Eine Lösungsmethode zur Behandlung nicht-linearer Randwertprobleme, ISNM 15 (1970), 245-257.

[103] Wacker, Hj., Ein Iterationsverfahren zur Lösung spezieller Randwertprobleme, Computing 8 (1972), 275-291.

[104] Wacker, Hj., Konstruktion von Iterationsketten und Behandlung von Verzweigungen, Meth. u. Verf. d. Math. Physik 6 (1972), 69-80.

[105] Wacker, Hj., Nichtlineare Homotopien zur Konstruktion von Startlösungen für Iterationsverfahren, in: Numerische Lösung nichtlinearer partieller Differential- und Integrodifferentialgleichungen, (eds. R. Ansorge - W. Törnig), Springer Lecture Notes 267 (1972), 51-67.

[106] Wacker, Hj., Übergangsmöglichkeiten zwischen verschiedenen Iterationsverfahren, ZAMM 4 (1974), T236-237.

[107] Wacker, Hj., Minimierung des Rechenaufwandes für spezielle Iterationsverfahren vom Typ Minimales Residuum, Computing 18 (1977), 209-224.

[108] Wacker, Hj., Zarzer, E., Zulehner, W., Optimal stepsize control for the globalized Newton method, Symp. "Imbedding Methods", Univ. Linz, Math. Inst. (1977)

[109] Wasserstrom, E., Numerical solutions by the continuation method, SIAM Review Vol. 15, Nr. 1 (1973), 89-119.

[110] Wendland, W., On the Imbedding Method for semilinear first order elliptic systems and related Finite Element Methods, Symposium "Imbedding Methods for the solution of nonlinear problems", Univ. Linz, Math. Inst. (1977), 1-48.

[111] Yakovlev, M. N., The solution of systems of non-linear equations by a method of differentiation with respect to a parameter, USSR Comp. Math. a. Math. Physics 4 (1964), 198-203.

[112] Židkov, E. P., Puzynin, I. V., Application of a continuous analogue of Newton's method to the approximate solution of a nonlinear boundary value problem, Soviet Math. Dokt., Vol. 9, Nr. 3 (1968), 575-578.

Research Projects:

Projekt Fe 125/1 DFG (1973 - 75).

Scientific management: M. Feilmeier - Hj. Wacker;

co-authors: P. Lory, H. Scheuring

(I) Bericht 1 Preprint Inst. f. Angew. Math. TU München,
1-56 (1973)

(II) Bericht 2 Preprint Nr. 7, Math. Inst. Univ. Linz,
1-44 (1974)

(III) Bericht 3 Preprint Nr. 26, Math. Inst. Univ. Linz,
1-62 (1975)

Projekt Nr. 2811, Fond z. Förderung der wissenschaftlichen Forschung (Austria) (1975 - 78)

Scientific management: Hj. Wacker;

co-authors: H. Brunner, F. J. Drexler, H. Engl, H. Hackl,
M. Traunmüller (MTA), E. Zarzer, W. Zulehner

(IV) Bericht 1 Preprint Nr. 65, Math. Inst. Univ. Linz,
1-126 (1976)

(V) Bericht 2 Preprint Nr. 77, Math. Inst. Univ. Linz,
1-157 (1977)

(VI) Bericht 3 Preprint Nr. 102, Math. Inst. Univ. Linz,
1-90 (1978)

THE TOPOLOGICAL THEORY OF AN EMBEDDING METHOD

J. C. Alexander

Department of Mathematics
University of Maryland
College Park, Maryland

The purpose of this report is to describe an embedding method which is suggested by results from algebraic and differential topology. It is the abstraction and generalization of a method that was developed to numerically locate fixed points. Only early forms of the method have been numerically implemented. Thus I do not have any actual numerical experience to report. Neither is any explicit algorithm developed, although the general form an algorithm should take is indicated. Rather the purpose of this report is to present an abstract program for designing algorithms and to show by examples that

[1]*Present address: University of Bonn, Western Germany*

Partially supported by NSF contracts and by SFB 4o, Bonn

ISBN 0-12-729250-0

the program is potentially useful for a number of interesting problems, including for example, one- and multi-parameter bifurcation problems. Proofs and other technical aspects of the method have appeared elsewhere , so we will be more informal and descriptive here and refer the reader to these other articles for more details. Also we will not work in complete generality. For example we assume all manifolds and functions are smooth (= C^{∞}). That amount of differentiability is never needed and the method is also applicable to the piecewise linear and simplicial situations. Indeed, one can trace some of the topological invariants used back to simplicial methods developed by B.C. Eaves and H. Scarf for finding fixed points. It will be shown in an appendix that the method is applicable to piecewise differentiable cases.

SOME PRECURSORS

Before developing the method itself, let us briefly consider some immediate predecesors. We mention only those methods that had a direct impact on the present formulation and refer the reader to the mentioned articles for further bibliography.

While studying M. Hirsch's proof of the Brouwer Fixed Point Theorem in John Milnor's *Topology From the Differentiable Viewpoint*, R.B.Kellogg, T.Y. Li and J.A. Yorke realized it could be made "constructive" in some sense and that it could be used to locate numerically fixed points of differentiable mappings of the n-dimensional disk D^n to itself. They reported their method and some numerical experimentation in

1974 [K-L-Y].

Let $g: D^n \to D^n$ be a smooth map and suppose g has no fixed points on the boundary S^{n-1} of D^n. Let C (interior D^n) be the set of fixed points. Define $h: D^n - C \to S^{n-1}$ as indicated in Figure 1. Since g is smooth, so is h as a map from an n-dimensional manifold to an (n-1)-dimensional manifold. The map h is of course the one used to change the question of fixed points into one about retracting D^n onto S^{n-1}. Now suppose that $x^o \in S^{n-1}$ is a *regular point* of h. That is, the n by (n-1) matrix of partials of h is of rank (n-1) at any point in $h^{-1}(x^o)$. Then by the implicit function theorem, $h^{-1}(x^o)$ is a one-dimensional manifold with x^o as its only boundary point. In particular, the other end of the curve in $h^{-1}(x^o)$ emanating from x^o must approach C. See Figure 2. Thus there is a path that can be followed from the boundary S^{n-1}, where everything is known, to the fixed point set C. The existence of a regular value x^o is guaranteed by Sard's Theorem which states that the set R of regular values is *residual* in S^{n-1}. That is, R is the countable intersection of open dense sets. Thus the complement of R is of measure zero in S^{n-1} and R is dense.[2]

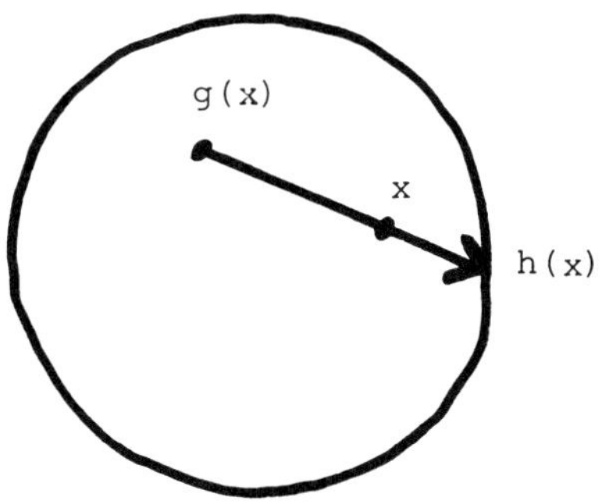

FIGURE 1. Defining $h: D^n - C \to S^{n-1}$

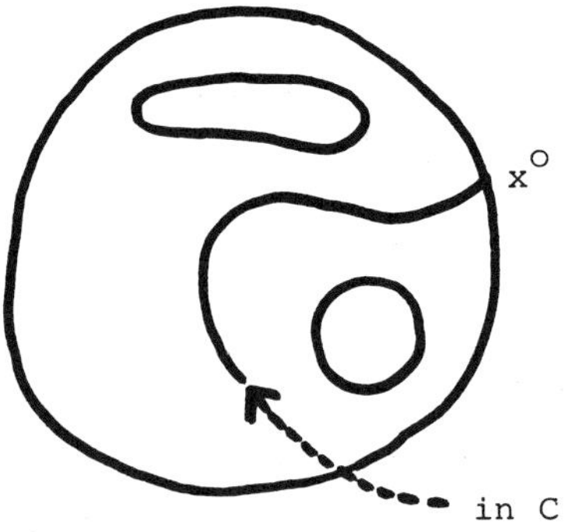

FIGURE 2. The inverse image $h^{-1}(x^o)$

Both probabilistically and topologically, residual sets are big. The algorithm takes the following form. Choose a random point in S^{n-1}, say x^o. Follow $h^{-1}(x^o)$ by whatever curve-following method seems best (Kellogg, Li, Yorke used Euler's method). If x^o is a regular value, $h^{-1}(x^o)$ is a curve and one eventually gets arbitrarily close to C. If $h^{-1}(x^o)$ ceases to be a manifold at some point (a probability zero occurance), go back and choose another x^o.

Kellogg, Li and Yorke also gave conditions (on the eigenvalues of g') that guarantee *all* $x^o \in S^{n-1}$ are regular. For this case they explicitly developed the equations for following the curve. As C is approached, the equations become those of Newton's method for finding a zero of x-g(x). Thus the method can be considered a globalization of Newton's method.

Subsequently, Stephen Smale [S] , who was interested in finding equilibria of economic systems, developed the following method for finding zeroes of functions $f: D^n \to R^n$. Assume f has no zeroes on S^{n-1} and obeys the following normal condition: the outward normal n_x to D^n at all $x \in S^{n-1}$ is mapped by the derivative f_* to a vector $f_* n_x$ such that the inner product $(f_* n_x, r_{f(x)})$ is never zero. Here $r_{f(x)}$ is the ray through $f(x)$ (see Figure 3). Let C equal $f^{-1}(o)$ and define $g: D^n - C \to S^{n-1}$ by $g(x) = f(x)/|f(x)|$. As before, by Sard's Theorem, a residual set of $y \in S^{n-1}$ is regular. Then $g^{-1}(y)$ is a smooth curve. The normal condition ensures that $g^{-1}(y)$ intersects $S^{n-1} = \partial D^n$ exactly once and thus the component of that point of intersection leads to a zero of f. Also Smale defined a differential equation on $g^{-1}(y)$ that guarantees that the zero will be reached in a finite amount of time. A feature of Smale's method that is made quite explicit is that while following $f^{-1}(y)$, the value of $|f(x)|$ does not necessarily decrease monotonically to zero (see Figure 4). It is not a method of steepest descent. It is necessary sometimes to allow f(x) to increase sometimes, to "go uphill".

Subsequently to this, Yorke recast the method as an embedding method. For example, consider again the problem of finding a zero of $f: D^n \to R^n$. Let $f_\alpha(x) = x - \alpha$ for any $\alpha \in D^n$,

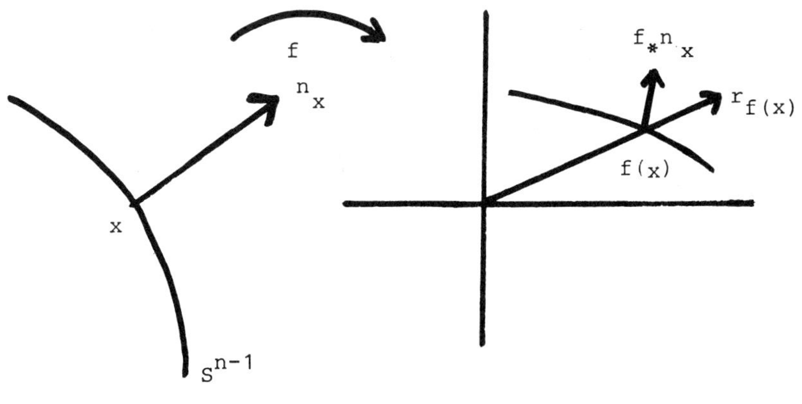

FIGURE 3. $(f_* n_x, r_{f(x)}) > o$

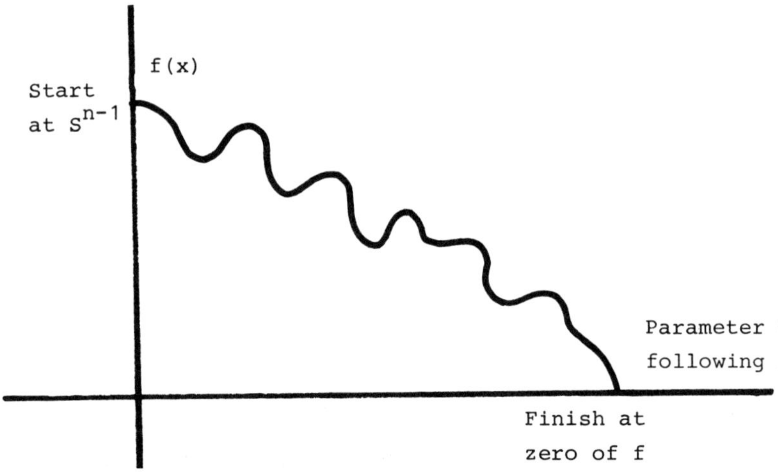

FIGURE 4.

let I be the unit interval [o,1] and define $F_\alpha : D^n \times I \to R^n$ by $F_\alpha(x,t) = (1-t)f_\alpha(x) + tf(x) = f(x) + (1-t)(f_\alpha(x) - f(x))$. Clearly $f_\alpha^{-1}(o) = \{\alpha\}$ and the idea is to follow the curve from α at t=o to the desired $f^{-1}(o)$ at t=1 (see Figure 5).

As before, we need that $F^{-1}(o)$ is a manifold. For this we use a standard consequence of Sard's Theorem (called the

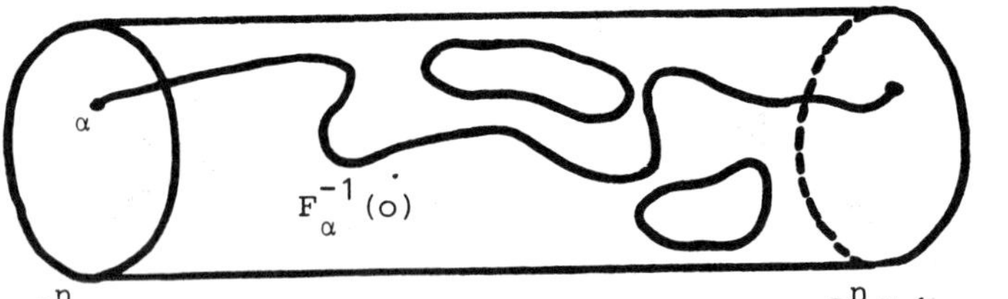

FIGURE 5. The embedding formulation

parameterized Sard's Theorem in [C-MP-Y]) that guarantees that for a residual set of α, the set $F_\alpha^{-1}(o)$ is a manifold over $[o,1)$. Since we have no control over $f(x)$, we have no control over the slice at $t=1$ of $F_\alpha^{-1}(o)$. We also need that $F_\alpha^{-1}(o)$ does not "escape" out the boundary $S^{n-1}\times I$. If for example $f(x)=x-g(x)$ for $g: D^n \to D^n$, then $F_\alpha^{-1}(o)$ does not escape and the curve must approach a fixed point of g at $t=1$.

Cast in this form, the method could be applied to a variety of other related problems. This formulation and such problems are the theme of [C-MP-Y]. The object of this report is to show how to generalize the method so it is applicable to a wide variety of problems.

THE FORMALISM

We present here the abstract, general formulation of the method. Afterwards some examples will be presented. For more details, somewhat more generality, and more examples, the reader is referred to [A-Y] .

Let M be a smooth m-dimensional manifold with boundary ∂M. Let $f: M \to R^n$ be a smooth map from M to Euclidean spaces and P a p-dimensional closed manifold without boundary in R^n. We want to "solve" the equation $f(x) \in P$.

Often $P=\{o\}$ and we are locating the zeroes of f. We assume $f(\partial M) \cap P = \emptyset$ so there are no solutions on the boundary of M. We will have a set $f_\alpha : M \to R^n$ of "initial" functions parameterized by $\alpha \in A$ for which we know $f_\alpha^{-1}(P)$. We define $F_\alpha : M \times I \to R^n$ by $F_\alpha(x,t)=(1-t)f_\alpha(x)+tf(x)$.

By a priori estimates we will know $F_\alpha(\partial M \times I) \cap P = \emptyset$.
The object is to "follow" $F_\alpha^{-1}(P)$ from the initial known $f_\alpha^{-1}(P)$ to the desired $f^{-1}(P)$. There are three problems we must confront:

(1) how to guarantee $F_\alpha^{-1}(P)$ connects $M_o = M \times 0$ and $M_1 = M \times 1$ so that we can get from one end to the other

(2) how to guarantee that $F_\alpha^{-1}(P)$ is nice-precisely, how to guarantee for a big set of α, that $F_\alpha^{-1}(P)$ is a manifold over $[o,1)$

(3) how to numerically follow $F_\alpha^{-1}(P)$.

We will give reasonably satisfactory answers to the first two. The third is work yet to be done.

First we reproduce a simple example from [A-Y], based on a suggestion of John Guckenheimer, that shows that $F_\alpha^{-1}(P)$ does not always connect M_o and M_1, even when it is impossible to

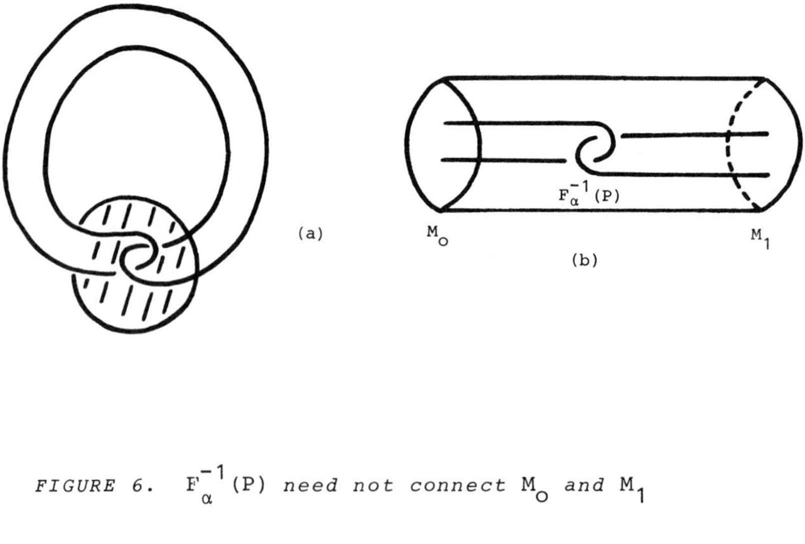

FIGURE 6. $F_\alpha^{-1}(P)$ *need not connect* M_o *and* M_1

deform f off of P. Let $P \subset R^3$ be the curve indicated in Figure 6(a), and let M be a two-dimensional disk D^2 with boundary S^1. Suppose f maps M so S^1 links P and M cuts across the right hand loop of P. Let f_α map M so it cuts across the left hand loop of P. (As α varies the vertical position of $f_\alpha(M)$ varies). Figure 6(b) shows how $F_\alpha^{-1}(P)$ is embedded in $M \times I$. It clearly does not connect M_0 and M_1.

Thus we can ask what reasonable invariant or condition guarantees that $F_\alpha^{-1}(P)$ connects M_0 and M_1. The above example shows that the homotopy class of ∂M in $R^n - P$ is not a suitable invariant. Jean Mallet-Paret has pointed out that Lusternik-Schnirelman category (which for example, detects zeroes of gradient functions) also is not. There exists an $f: S^1 \to R$ and a homotopy f^t of f such that the zeroes of ∇f^t do not connect the two ends of the cylinder $S^1 \times I$ of the homotopy. Such an f is exhibited in Figure 7. The standard S^1 in R^3 is deformed as indicated. The function f is the height function, the projection to the vertical axis. The zeroes of the gradient are the maxima and minima.

(1) (2) (3) (4) (5) (6) (7)

FIGURE 7.

The following condition however, suffices for a large class of problems: *if*

$$f_*: h_*(M,\partial M) \longrightarrow h_*(R^n,R^n-P)$$

is non-zero for some (generalized) homology theory h_*, *then* $F_\alpha^{-1}(P)$ *connects* M_o *and* M_1. (The dual cohomological condition also works).

This result does not depend on any particular form of f_α or F_α, on the smoothness of f, or even on the fact that M,R^n, and P are manifolds. Thus it is applicable to other versions of the method. The condition may look alien to a numerical analyst. Fortunately it does not bear at all on the actual numerical programs. This aspect can be handled separately and taken on faith by anyone interested in carrying out the numerical analysis. Moreover it is a natural generalization of previous conditions. The existence of an element in $f^{-1}(P)$ is often established by some invariant that comes from the condition.

For example, suppose M is an n-dimensional connected manifold and $P=\{o\}$. Let h_* be ordinary homology H_*. Then $H_*(M,\partial M)$ and $H_*(R^n,R^n-\{o\})$ are both copies of the integers and f_* is multiplication by some integer δ. This integer δ is nothing but the classical degree of f. Thus *if the degree of f is non-zero, the condition holds*. For example, consider Smale's method. The normal condition implies the degree of f is ± 1 and thus the homological condition is satisfied. Similarly, the fixed point index of $g: D^n \to D^n$ is the degree of $f(x)=x-g(x)$ with $P=\{o\}$, and equals 1 for a map of D^n to itself. All the applications of [C-MP-Y] involve degree.

Another word about faith: in standard embedding problems, the set to be followed is parameterized by the homotopy variable. Geometrically, one always goes from left to right. If the set turns to the left, the effort is abandoned. Here the set may well turn to the left temporarily. This is exactly analogous to $|f(x)|$ not decreasing monotonically in Smale's method. Rather than abandon the effort at such a point, with our faith in topology, we proceed onward, guaranteed that eventually we will reach our goal (Figure 8).

Let us now consider the second problem. Call a parameter *acceptable* if $F_\alpha^{-1}(P)$ is a manifold over $[o,1)$. Call it *completely acceptable* if $F_\alpha^{-1}(P)$ is a manifold over all $[o,1]$, and for $\varepsilon > o$, call α' ε-*acceptable* if $F_\alpha^{-1}(P)$ is a manifold over $[o,1-\varepsilon]$. Let A be a smooth manifold without boundary (the *parameter manifold*) and let $\phi: A \times M \to R^n$ be a smooth mapping. For each $\alpha \in A$, define $f_\alpha: M \to R^n$ by $f_\alpha(x)=\phi(\alpha,x)$. Similarly define $\Phi: A \times M \times I \to R^n$ by $\Phi(\alpha,x,t)=(1-t)\phi(\alpha,x)+tf(x)$, and $F_\alpha: M \times I \to R^n$ by $F_\alpha(x,t)=\Phi(\alpha,x,t)=(1-t)f_\alpha(x)+tf(x)$. We require ϕ to be such that $\Phi(A \times \partial M \times I) \cap P=\emptyset$. Call ϕ *sufficient* if for each $x \in M-\partial M$, the map $A \to R^n$ defined by

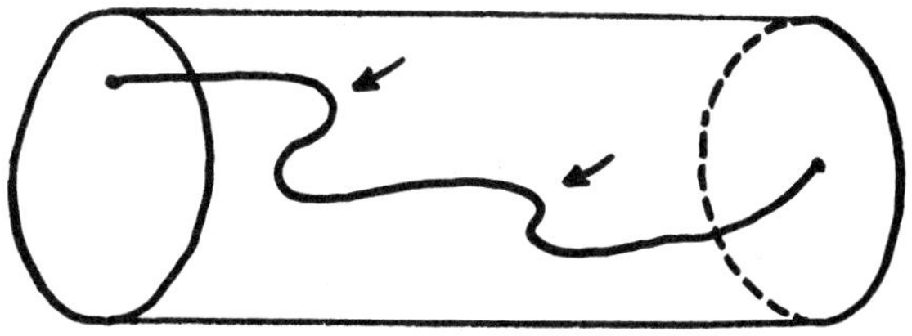

FIGURE 8. Have faith !

$\alpha \to \phi(\alpha, x)$ is a submersion. That is, for each x, the matrix of partials $(\partial\emptyset/\partial\alpha)$ has rank n at each point. The seminal example is $M=D^n$, A=interior D^n, and $\phi(x,\alpha)=x-\alpha$.

The major result is: *if* $\Phi: A \times M \to R^n$ *is sufficient, the* α *in a residual subset of* A *are acceptable*.

The proof simply amounts to remarking that $\Phi: A \times M \times I \to R^n$ is transverse to P, at least over $[0,1)$. The proof is then a standard consequence of a basic result about transversality. (Briefly, consider $W=\Phi^{-1}(P) \subset A \times M \times I$. By transversality, W is a manifold. Let $\pi: W \to A$ be the restriction of the projection. One observes that if $\alpha \in A$ is a regular value of π, then α is acceptable. By Sard's Theorem, the set of such α is residual). Thus we turn the α-transversality for all x into x-transversality for almost all α.

It should be noted that when α is acceptable, the manifold $F_\alpha^{-1}(P)$ need not be one-dimensional. In fact dim $f_\alpha^{-1}(P)=m-(n-p)$ and dim $F_\alpha^{-1}(P)=(m+1)-(n-p)$, so we may very well be faced with the task of following a higher dimensional manifold from M_0 to M_1 (Figure 9). Indeed we will consider problems of interest

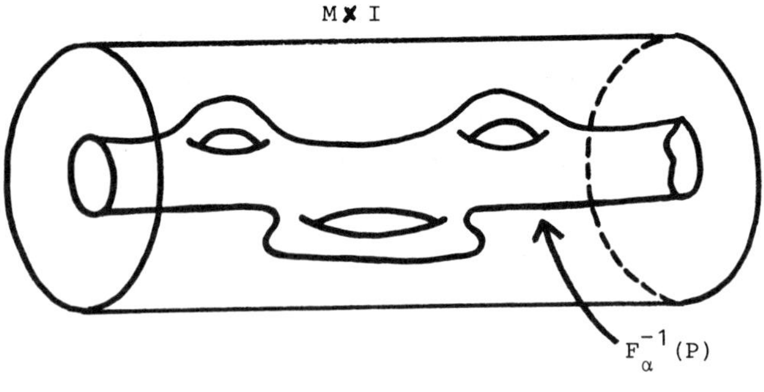

FIGURE 9. $F_\alpha^{-1}(P)$ *may have dimension greater than one*

with dim $F_\alpha^{-1}(P) > 1$. Let us briefly and informally turn to the third problem and consider the form a numerical algorithm should take. Since P is a submanifold of R^n, it comes equipped with a tubular neighbourhood N in R^n. For acceptable α , not only is P pulled back by F_α to a submanifold $F_\alpha^{-1}(P)$ in $M \times I$ (at least over $[o,1))$, but some N is pulled back by F_α to a tubular neighbourhood of $F_\alpha^{-1}(P)$. That is, near $F_\alpha^{-1}(P)$ we have a neighbourhood with a bundle structure, so near $F_\alpha^{-1}(P)$ we know how to get back to $F_\alpha^{-1}(P)$. In other words we can correct small errors. The idea then is to trace out all of (a component of) $F_\alpha^{-1}(P)$ starting from $f_\alpha^{-1}(P)$. Suppose dim $F_\alpha^{-1}(P)=d$. Suppose we have reached some point $x \in F_\alpha^{-1}(P)$ (see Figure 1o). We calculate the d tangent directions and the normal directions to $F_\alpha^{-1}(P)$ at x. We take a step in each of the d tangent directions, ending up say at $x_1',\ldots,x_d'$. Calculation the $F_\alpha(x_i')$ and using the normal directions at x, we correct the x_i' to x_i. If necessary more corrections can be made.

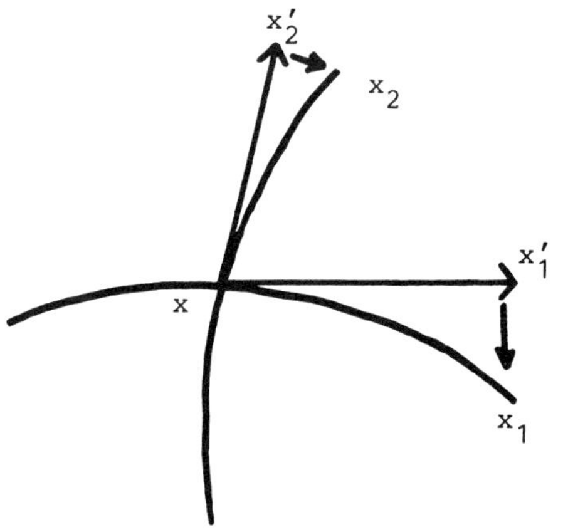

FIGURE 1o. The iterative step

One needs a method to recognize when the process has gone around a bend or through a hole and is back on part of $F_\alpha^{-1}(P)$ already found. Calculating the tangent and normal directions involves calculating partial derivatives of F_α. The matrix of partials will have rank d at each point for acceptable α. Put the other way round, if the matrix of partials has rank less than d, α is unacceptable and a new α must be chosen.

Note that the process involves doing the same calculations on d (or possibly more) different points. This could possibly be made efficient on an "array" computer-one that has the capability to perform the same operation on the values of a large number of memory positions at one time.

EXAMPLES

1. Fixed Point Problems. This has already been considered in detail. We might mention the theory holds for mappings in Banach spaces that have fixed point indices. In particular A-proper mappings should be amenable to the theory of this report. Moreover, the method will work on manifolds M more general than a disk D^n. The requirement is that the map f of interest have non-zero Lefschetz number.

2. Zeroes of polynomials.[3] Let $p(z)=z^n+a_{n-1}z^{n-1}+\ldots+a_o$ be a polynomial with complex coefficients. Let $M=D^2$ be a disk in the complex plane C so large that all roots of p lie inside D^2 (or extend p to a map of the Riemann sphere S^2 and work on S^2). As a map $p: (D^2, \partial D^2) \to (C, C-\{o\})$ has degree n and thus the method will work. Let $A=(\text{interior}(D^2)^n - \text{diagonal})$ and if $\alpha=(z_1,\ldots,z_n)$, let $\phi(\alpha,z)=p_\alpha(z)=(z-z_1)\ldots(z-z_n)$. Then $\partial p_\alpha(z)/\partial z_i=-(z-z_1)\ldots(z-z_{i-1})(z-z_{i+1})\ldots(z-z_n)$ and ϕ is sufficient. Once $\alpha=(z_1,\ldots,z_n)$ is fixed, one gets n paths emanating from the n different $z_1,\ldots,z_n$ and in this case, local degree considerations guarantee that the n different paths lead to the n roots (possibly multiple) of p(z) (see Figure 11). The paths do not intersect or amalgamate except possibly at the very end at t=1.

3. Inverse Eigenvalue Problems. The additive inverse eigenvalue problem is: given an $(n \times n)$ complex matrix m and n complex numbers $(\xi_1,\ldots,\xi_n)$ (possibly with repetitions), find a diagonal matrix Δ with entries $(d_1,\ldots,d_n)$ such that

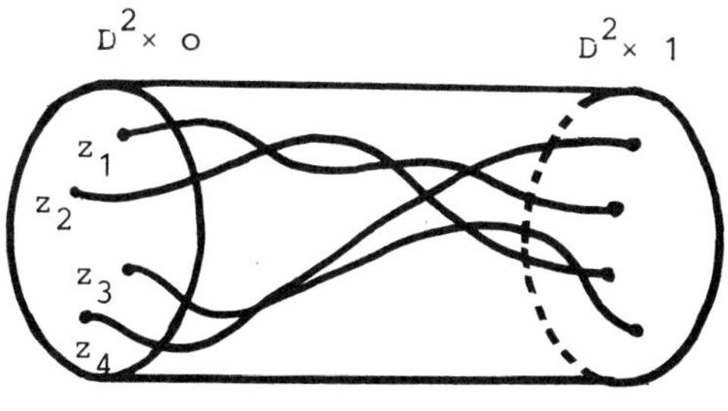

FIGURE 11. n paths from $z_1,\ldots,z_n$ to the n roots of p(z)

the eigenvalues of $m+\Delta$ are precisely $(\xi_1,\ldots,\xi_n)$. It is always possible to do so, generally in n! different ways.

Let $\bar{a}_{n-1},\ldots,\bar{a}_o$ be the coefficients of

$$r^n-\bar{a}_{n-1}r^{n-1}+\ \ldots\ +(-1)^n\bar{a}_o = (r-\xi_1)\ \ldots\ (r-\xi_n)$$

Define $f(\Delta)=(a_{n-1},\ldots,a_o)\ \varepsilon\ C^n$ if: the characteristic polynomial of $m+\Delta$ equals

$$r^n-a_{n-1}r^{n-1}+\ \ldots\ +(-1)^n a_o$$

Consider $\Delta\ \varepsilon\ C^n$ and let $M=D^{2n-1}\subset C^n$ be a large disk (or extend to M = complex projective space CP^n), and P be the point $(\bar{a}_{n-1},\ldots,\bar{a}_o)\ \varepsilon\ C^n$. Then $f:(D^{2n-1},\partial D^{2n-1}) \to (C^n,C^n-P)$ has degree n!. Thus M_1 and M_1 are connected by any $F_\alpha^{-1}(P)$ in $M\times I$. Let $A=C^{n^2}$ be the space of $(n\times n)$ matrices and $\phi:A\times M\to C^n$ be $\phi(\alpha,\Delta)$ = coeff. of char. polyn. of $(\alpha+\Delta)$. The reader is invited to investigate smaller dimensional sufficient A.

4. *Bifurcation.* Let $F:R^k\times R^m\to R^m$ be smooth. Let λ denote the variable in R^k which we consider a parameter, and let x denote the variable in R^m. Suppose $f(\lambda,o)=o$ for all λ and that the x-derivative $D_x f(\lambda,x)|_{x=o}$ (which is a linear map from R^m to R^m for each λ) is non-singular for $\lambda\neq o$. Then $\lambda=o$ is a possible bifurcation point for the zeroes of f.

There is an invariant γ_f defined, which takes values in an abelian group π^s_{k-1} and if $\gamma_f\neq o$, global bifurcation of $f^{-1}(o)$ is guaranteed. For example, if k=1, the group π^s_{k-1} is the integers and γ_f is the difference of the degrees of

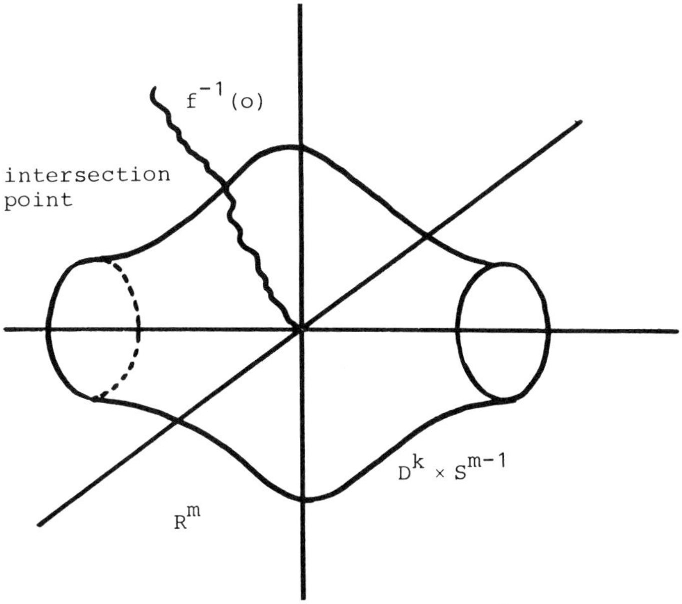

FIGURE 12. Locating bifurcation of $f^{-1}(o)$

$D_x f(\lambda, x)\big|_{x=o}$ between $\lambda > o$ and $\lambda < o$. Thus we have the non-linear eigenvalue result of Rabinowitz. If k=2, we can apply the result to the complex eigenvalue problem considered by Ize or the Hopf bifurcation result of Alexander-Yorke. If we replace R^m by Banach space, the result still holds if we put some restriction on f, e.g. A-proper.

What we observe here is that if $\gamma_f \neq o$, the method of the paper offers a way to numerically locate $f^{-1}(o)$. Let S^{k-1} be a sphere in R^k centered at o, and let S^{m-1} be a small sphere in R^m centered at the origin. Then $S^{k-1} \times S^{m-1} \subset R^k \times R^m$ does not intersect $f^{-1}(o)$. Let D^k be a k-dimensional disk and let $D^k \times S^{m-1}$ be located in $R^k \times R^m$ so that the boundary of $D^k \times S^{m-1}$ is the above $S^{k-1} \times S^{m-1}$. See Figure 12. It is not necessary that $D^k \times S^{m-1}$ sit in $R^k \times R^m$ in the standard way. Any way will do as long as the boundary sits correctly.

The theory guarantees that if $\gamma_f \neq o$, then $f^{-1}(o)$ intersects $D^k \times S^{m-1}$. Moreover the theory of the present paper guarantees that if $\gamma_f \neq o$, the method of this paper can be applied to locate that intersection. Also if f satisfies a non-degeneracy condition (to wit: there exists $\varepsilon > o$ such that $|D_x f(\lambda,x)|_{x=o} \cdot u| \geq \varepsilon|\lambda||u|$ for λ inside S^{k-1}, $u \in R^m$), then there are good bilinear initial f_α constructed out of "non-degenerate multiplications of R^k on R^m" or skewed Clifford algebra actions.

Let us note the following points. During the homotopy from f_α at t=o to f at t=1, we do not vary $M=D^k \times S^{m-1}$, but rather the function f. Thus we are not "following the bifurcation branch", but rather finding its intersection with preassigned cylinders. Also the method does not care about the stability of the branch. Moreover the global character of the bifurcation branch is accessible; the method is not just local. Also note there is no restriction on k. The method offers an approach to multi-parameter bifurcation.

This example is interesting for the general theory of this paper also. If k=1, the degree detects the bifurcation. However, if $k > 1$, ordinary homology definitely does not suffice. A generalized homology (stable homotopy) must be used. Also the sets $F_\alpha^{-1}(o)$ are k-dimensional. Thus for example for complex nonlinear eigenvalue problems or Hopf bifurcation, the manifolds $F_\alpha^{-1}(o)$ are surfaces. This example shows the generality we have developed is not idle.

We might mention that Chow, Mallet-Paret and Yorke have recently proved a global Hopf bifurcation theorem more general than that mentioned above. They use Fuller degree rather than the invariant above. There is an open problem from the point of view of this paper; it is not known if non-zero Fuller degree guarantees the solutions of the problem through the homotopy connect the two ends of the cylinder.

5. *The Hot-Dog Problem.* We consider a geometric problem for which $P \neq \{o\}$ (although to be sure it reduces to one with $P = \{o\}$ easily enough). Let C be any compact body in R^3. The hot-dog problem is to find a cube that circumscribes C. That is, C is contained in the cube and each of the six faces of the cube touches C. The problem got its name from the fact that a hot-dog is a test body. For history and references the reader is referred to the section "Theorems of Kakutani, Dyson, and Yamabe-Yujobo" in N. Steenrod's *Reviews of Papers in Algebraic and Differential Topology.*

Let SO(3) be the space of 3 mutually perpendicular unit vectors v_1, v_2, v_3 in R^3 based at the origin, ordered so as to induce the standard orientation on R^3. Consider the rectangular solid with edges pointing in the directions v_1, v_2, v_3 that circumscribes C. Let the lengths of the edges be l_1, l_2, l_3, respectively. Define $f: SO(3) \to R^3$ by

$f(v_1,v_2,v_3) = (l_1,l_2,l_3)$. The function f is continuous and is smooth if C is smooth and strictly convex (if C is not strictly convex, f is only piecewise smooth). If the image of f intersects the diagonal of R^3, the rectangular solid is a cube.

Thus we let M=SO(3), P=the diagonal in R^3. Not only does the circumscribing cube always exist , but the methods of the present paper can be used to find it. We define f_α by letting C_α be a thin needle. The parameter α describes its orientation in R^3. The circumscribung cube sits so C_α lies diagonally across it. Note that $f_\alpha^{-1}(P)$ is one-dimensional since the circumscribing cube has one degree of (rotational) freedom. Thus $f_\alpha^{-1}(P)$ is two dimensional.

Although this example is somewhat frivolous, there is a serious aspect. C.T. Yang gave a unified proof of this problem and other related problems using an invariant he defined. E. Fadell and P. Rabinowitz have generalized Yang's invariant and proved such things as bifurcation results for Hamiltonian systems. Again there is an open problem since it is not known if non-zero Yang invariant guarantees $F_\alpha^{-1}(P)$ connects M_o and M_1.

SOME COMMENTS

The formulation we have presented is for smooth functions. So much differentiability is not needed, although presumably the more differentiability available, the better the numerical methods will work, as is the case for example with differential equations. For the theory only the amount called for by Sard's Theorem is required (usually one more degree of differentiability than the dimension of $F_\alpha^{-1}(P)$). In any event the problem must be made discrete to use numerical methods. In [C-MP-Y] for example, where $F_\alpha^{-1}(P)$ is one-dimensional, there is a first-order differential equation defined in a neighbourhood of $F_\alpha^{-1}(P)$ for which $F_\alpha^{-1}(P)$ is an integral. The numerical techniques for integrating differential equations can be used. We have indicated how to attack the higher-dimensional cases.

That the problem starts out continuous and must be made discrete is not necessarily a disadvantage. One is not committed to any preassigned triangulation as is the case in some simplicial methods. Moreover with some information about the second partials of F_α at $x \varepsilon F_\alpha^{-1}(P)$, the curvature of $F_\alpha^{-1}(P)$ at x can be calculated. If the curvature is large, smaller iterative steps can be taken to "go around the bend" more carefully. Li has investigated this in the fixed point problem.

Finally, let us consider some aspects of choosing a "random" α. Of course any computing machine has only a finite number of numbers to work with. Even when idealized it has only a countable number, say the rationals or those with finite dyadic expansions. The complement of a countable set is residual. Thus the set of unacceptable α might be exactly the set the machine has available to it.

There are several things that can be noted. If f itself is transverse to P the set of completely acceptable α is not only residual but open. (For example, in the fixed point problem, if the fixed points are non-degenerate, f is transverse to P. In example 2, if the roots of the polynomial are distinct, f is transverse to P). In any open dense set there are lots of rationals. Indeed one would be very unfortunate to choose an unacceptable α. Even if f is not transverse to P, for any $\varepsilon > o$ the set of ε-acceptable α is open and dense if $\emptyset$ is sufficient.

It would be possible to impose conditions on f so that all (or some preassigned part of) α are acceptable. This was done by Kellogg, Li and Yorke. However this would severely limit the problems to which the method is applicable. On the other hand, it might be possible to follow $F_\alpha^{-1}(P)$ even if it is not a manifold at points. For example one might reasonably hope that if M,P,A, and ϕ are algebraic (i.e. given by polynomials), that $F_\alpha^{-1}(P)$, although it might have cusps and nodes and such things, would still be reasonable enough to follow. In this regard I offer the conjecture that if M,P,A and ϕ are algebraic and ϕ is sufficient, then the set of acceptable α is always open.[4]

Even for the general case it might prove worthwhile to continue if $F_\alpha^{-1}(P)$ ceases to be a manifold. The algorithm could simply search in all directions from the last point to find a new point of $F_\alpha^{-1}(P)$. Indeed the requirement of having $F_\alpha^{-1}(P)$ be a manifold could be bypassed completely by just searching at each iterative step. Then no partials need be computed and the parameter manifold A need not be used.

I feel the abstract theory of the method is rather complete. What is needed now is the development and implementation of algorithms with the concomitant error and convergence analyses to see if in fact the method is practical for any interesting problems.

APPENDIX I

PIECEWISE DIFFERENTIABLE FUNCTIONS

If $f:M \to R^n$ is not smooth on all of M, but rather M is divided into pieces such that f is smooth on each piece, the results of this report still hold. Here we present an outline of this generalization. This generalization is the result of conversations with Jim Yorke.

We discuss only the case dim M = dim R^n = n and P={o} and leave other generalizations to the interested reader. We suppose M is broken up into pieces such that the boundaries of the pieces are submanifolds of M. More precisely, let $U_1,\ldots,U_I$ be a finite open cover of M. Suppose for each $i=1,\ldots,I$, there are defined smooth functions

$$\Psi_{i,1},\ldots\ldots,\Psi_{i,J} : U_i \to R$$

where J=J(i), which satisfies the following transversality

conditions:

i) O is a regular value of each $\Psi_{i,j}: U_i \to R$

ii) O is a regular value of each $\Psi_{i,j} \times \Psi_{i,k} : U_i \to R \times R$ if $j \neq k$. Here $(\Psi_{i,j} \times \Psi_{i,k})(x) = (\Psi_{i,j}(x), \Psi_{i,k}(x))$.

(Recall that O is a regular value of a function $\Psi: U \to R^m$ if the matrix of partials of Ψ at any point $x \in \Psi^{-1}(o)$ has rank m. It is the same thing to say Ψ is transversal to O). Condition i) means the $\Psi_{i,j}^{-1}(o)$ are codimension 1 submanifolds of M. Condition ii) means these submanifolds meet transversally, i.e. never at a zero angle. For $i=1,\ldots,I$ let $V_i = \bigcap_{j=1}^{J(i)} \{x \in M \mid \Psi_{i,j}(x) > o\}$. We require that $V_i \cap V_{i'} = \emptyset$ if $i \neq i'$ and that $M = \bigcup_i \bar{V}_i$. Such a collection of data we call a *piecewise smooth decomposition* of M. The *boundaries* or *edges* of V_i consist of those x such that some $\Psi_{i,j}(x) = o$. The *corners* of V_i are those sets where $\Psi_{i,j}(x) = \Psi_{i,k}(x) = o$ for some $j \neq k$. The V_i themselves are the *regions* or *faces* of the decomposition. See Figure I1.

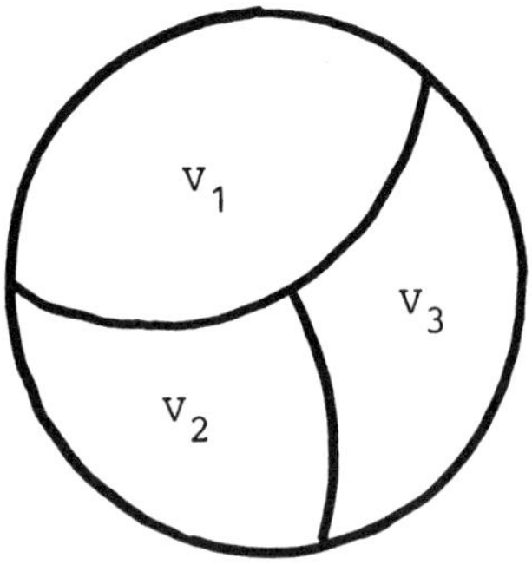

FIGURE I1. A piecewise smooth decomposition

A continuous function $f:M \to R^n$ is called *piecewise smooth* if there is a collection of smooth functions $f_i:U_i \to R$ for $i=1,...,I$ such that $f|U_i = f_i$. Intuitively, f is smooth on each region V_i but can have sharp turns at the boundaries. Let $\pi:M \times I \to M$ be the projection. We define a piecewise smooth decomposition of $M \times I$ by letting $\tilde{U}_i = \pi^{-1}U_i$, $\tilde{\Psi}_{i,j} = \Psi_{i,j}\pi$. Thus $M \times I$ is decomposed by cylinders $V_i \times I$. See Figure I2. Now suppose $\phi:A \times M \to R^n$ is a smooth function and let $f_\alpha(x) = \phi(\alpha, x)$ for each $\alpha \in A$. Let $F_\alpha:M \times I \to R^n$ be defined by $F_\alpha(x,t) = (1-t)f_\alpha(x) + tf(x)$ as before. Similarly define $F_{i,\alpha}:V_i \times I \to R^n$ by $F_{i,\alpha}(x,t) = (1-t)f_\alpha(x) + tf_i(x)$. We say $F_\alpha^{-1}(o)$ is *transverse to the decomposition* of $M \times I$ if:

t-o) O is a regular value of each $F_{i,\alpha}: V_i \times I \to R^n$

t-i) O is a regular value of each $F_{i,\alpha} \times \tilde{\Psi}_{i,j}:U_i \times I \to R^n \times R$

where $(F_{i,\alpha} \times \tilde{\Psi}_{i,j})(x,t) = (F_{i,\alpha}(x,t), \Psi_{i,j}(x))$

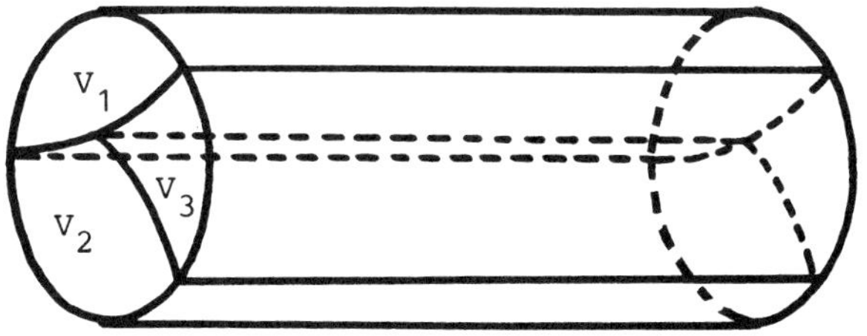

FIGURE I2. A piecewise smooth decomposition

t-ii) O is a regular value of each

$$F_{i,\alpha} \times \tilde{\Psi}_{i,j} \times \tilde{\Psi}_{i,k} : U_i \times I \to R^n \times R \times R \text{ for } j \neq k, \text{ where}$$

$$(F_{i,\alpha} \times \tilde{\Psi}_{i,j} \times \tilde{\Psi}_{i,k})(x,t) = (F_{i,\alpha}(x,t), \Psi_{i,j}(x), \Psi_{i,k}(x))$$

Condition t-o) implies that each $F_{i,\alpha}^{-1}(o)$ is a smooth curve and thus $F_\alpha^{-1}(o)$ is a piecewise smooth curve. Condition t-i) implies that each $F_{i,\alpha}^{-1}(o)$ intersects each boundary transversally; i.e. at non-zero angle. Condition t-ii) implies $F_\alpha^{-1}(o)$ contains no corner points.

If t-o,i-ii) hold except for t=1, we say $F_\alpha^{-1}(o)$ is transverse to the decomposition over $[o,1)$.

Intuitively, $F_\alpha^{-1}(o)$ is transverse to the decomposition if it is a smooth curve on each $\bar{V}_i \times I$, but it can have sharp bends when it crosses the boundaries. Moreover it never crosses at a corner.

We have the following result: *if* $\phi : A \times M \to R$ *is sufficient, the set* $F_\alpha^{-1}(o)$ *is transverse to the decomposition over* $[o,1)$ *for a residual set of* α. *For each* $\varepsilon > o$, *the set* $F_\alpha^{-1}(o)$ *is transverse to the decomposition over* $[o,1-\varepsilon]$ *for an open dense set of* α. *Moreover, if* f *is transverse to* O *and* $f^{-1}(o)$ *contains no corner points,* $F_\alpha^{-1}(o)$ *is transverse to the decomposition for an open dense set of* α.

The proof is rather trivial. We define $\Phi_i : A \times U_i \times I \to R^n$ by $\Phi_i(\alpha,x,t) = (1-t)\phi(\alpha,x) + tf_i(x)$ and note that each of the following is transverse to O (over $[o,1)$ or $[o,1]$ as the case may be): Φ_i itself, $\Phi_i \times \tilde{\Psi}_{i,j} : A \times U_i \times I \to R^n \times R$ where $(\Phi_i \times \tilde{\Psi}_{i,j})(\alpha,x,t) = (\Phi_i(\alpha,x,t), \Psi_{i,j}(x))$, and for $j \neq k$ $\Phi_i \times \tilde{\Psi}_{i,j} \times \tilde{\Psi}_{i,k} : A \times U_i \times I \to R^n \times R \times R$ where $(\Phi_i \times \tilde{\Psi}_{i,j} \times \tilde{\Psi}_{i,k})(\alpha,x,t) = (\Phi_i(\alpha,x,t), \Psi_{i,j}(x), \Psi_{i,k}(x))$. This immediately implies the result.

A number of generalizations, which we do not go into, immediately suggest themselves. We point out only that if $F_\alpha^{-1}(P)$ is to be of dimension greater than one, conditions (i,ii) and (t-o,i,ii) need to be augmented by conditions (iii,iv,...) and (t-iii,iv,...).

APPENDIX II

ALGEBRAIC DETAILS FOR EXAMPLES 4 AND 5

Examples 4 and 5 are new and the verifications that $F_\alpha^{-1}(P)$ connects M_o and M_1 are non-trivial. Although they do not bear on the numerical analysis aspects of the examples, we outline the verifications here for the sake of completeness.

To verify example 4 we need to show

$$f_*: h_*(D^k \times S^{m-1}, S^{k-1} \times S^{m-1}) \rightarrow h_*(R^m, R^m - \{o\})$$

is non-trivial. We let h_* be stable homotopy. Let $\sum_{k+m-1}$ be the class of $D^k \times S^{m-1}$ with the obvious trivial framing in $h_*(D^k \times S^{m-1}, S^{k-1} \times S^{m-1})$. The following diagram commutes:

$$\begin{array}{ccc} h_{k+m-1}(D^k \times S^{m-1}, S^{k-1} \times S^{m-1}) & \xrightarrow{f_*} & h_{k+m-1}(R^m, R^m - \{o\}) \\ \downarrow \partial & & \downarrow \partial \\ h_{k+m-2}(S^{k-1} \times S^{m-1}) & \xrightarrow{f_*} & h_{k+m-2}(R^m - \{o\}) = \pi^s_{k-1} \end{array}$$

Let $\partial\sum_{k+m-1} = \sigma_{k+m-2} \varepsilon h_{k+m-2}(S^{k-1} \times S^{m-1})$. It is shown in [A], [A-F] that $f_*\sigma_{k+m-2} = \gamma_f \varepsilon \pi^s_{k-1}$. Thus if γ_f is non-zero, neither is $f_*\sum_{k+m-1}$. This proves f_* is non-trivial.

To verify example 5, an extension of our homological condition is required. In fact $f_* : h_*(SO(3)) \to h_*(R^3, R^3-P)$ is always trivial (for any P, in fact). However there is symmetry to the map we can use. The cyclic group G_3 of three elements acts on SO(3) by permuting v_1, v_2, v_3 cyclically and on $R^3 = \{(x_1, x_2, x_3)\}$ by permuting the coordinates cyclically. Clearly the diagonal P is invariant and f is equivariant.

In general suppose a group G acts on the right on M and on R^n so as to leave P invariant, and suppose $f:(M,\partial M) \to (R^n, R^n-P)$ is G-equivariant. Let X be any left G-space. The extended condition can now be stated (for use below, we state it in terms of cohomology): if $f^* : h^*(R^n \times_G X, (R^n-P)\times_G X) \to h^*(M\times_G X, \partial M\times_G X)$ is non-trivial for some cohomology h^* and some X, then $F_\alpha^{-1}(P)$ connects M_o and M_1 in $M \times I$. We let $G=G_3$ and $X=S^3$, where S^3 is the unit sphere in two dimensional complex space: $S^3 = \{(z_1, z_2) \mid |z_1|^2 + |z_2|^2 = 1\}$. Let G_3 operate on S^3 by taking (z_1, z_2) to $(z_1 e^{2\pi i/3}, z_2 e^{2\pi i/3})$. The quotient space S^3/G_3 is a lens space L^3. Let h^* be ordinary cohomology H^*. Let R^2 be the subspace of R^3 perpendicular to P. There is an equivariant deformation-retraction $(R^3, R^3-P) \to (R^2, R^2-\{o\})$. The induced G_3 action on the plane R^2 is rotation by $2\pi/3$. Moreover $(R^2\times_G S^3, (R^2-\{o\})\times_G S^3)$ can be considered the Thom space of a bundle ξ over L^3. Also $SO(3)\times_G S^3$ is a bundle over L^3 with fiber SO(3).

From the commutative diagramm

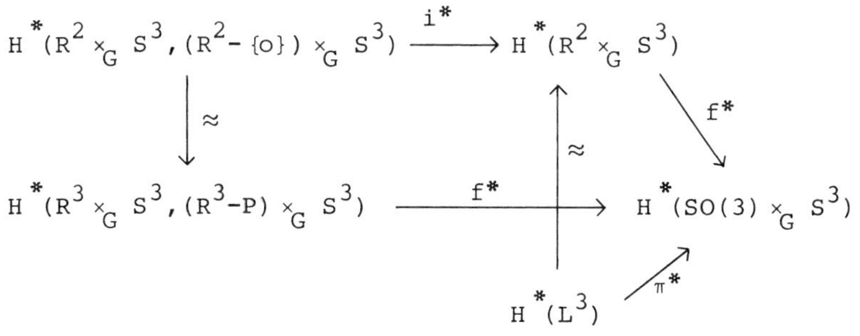

it can be seen that f^* is non-trivial if and only if $f^* i^*$ (Thom class of ξ) $\neq o$ if and only if π^* (Euler class of ξ) $\neq o$. The cohomology of L^3 is known ($H^o = Z$, $H^1 = o$, $H^2 = G_3$, $H^3 = Z$) and the Euler class of ξ is a generator of H^2.

As a space, SO(3) is three-dimensional real projective space, so $H^*(SO(3))$ is well-known. Also SO(3) is a group and G_3 acts as a subgroup. Therefore G_3 acts trivially on $H^*(SO(3))$. Using all this, the Serre spectral sequence for the fibration $SO(3) \times_G S^3 \to L^3$ can be seen to have no non-zero differentials and thus π^*(Euler class of ξ) $\neq o$. This establishes the result. It may be noted that the three directions need not be perpendicular. The method of proof also works for related problems; e.g. Dyson's theorem. On the other hand, the method of proof fails for higher dimensions.[5]

NOTES ADDED IN PROOF

[2]This is of course, not correct as stated. The complements of a residual (or even open dense) set can have positive measure. Call a set *full* if its complement has measure zero. Sard's Theorem yields a full, residual set. Throughout this paper, for residual read residual and full.

[3]T.Y. Li has programmed this particular example to see how well the method described in this paper works. Preliminary results indicate it works quite well if the polynomial has no multiple roots. If there are multiple roots the tubular neighbourhoods involved shrink as t nears 1 and the method becomes very slow near t=1.

[4]The conjecture is true. Indeed the set of unacceptable α are contained in a proper algebraic subset of A. This can be proved by emulating the proof of the differentiable case, but using the "algebraic Sard's Theorem". (D. Mumford, Algebraic Geometry I, Complex Projective Varieties, Springer Verlag, 1976, p.42).

[5]There is another condition which guarantees that $F^{-1}(P)$ connects M_o and M_1 in $M \times I$. It seems to be stronger if G is sufficiently complicated. The idea goes back to a result of A. Heller, On equivariant maps of spaces with operators, Ann. of Math. <u>55</u> (1952), 223-231, § 3.3.3.4. For convenience suppose $\partial M = \emptyset$. Let $\pi : (M \times (N-P))/G \to M/G$ be the map induced by π

in any homology (cohomology) has a right (left) inverse. The inverse is an $h^*(M/G)$-module homomorphism and respects stable operations. This can be shown by assuming $F^{-1}(P)$ splits into closed disjoint G-invariant subsets Q_o and Q_1 with $Q_o \cap M_1 = Q_1 \cap M_o = \emptyset$, and then splitting in two ways the Mayer-Vietoris sequence for $(M \times I-Q_o)/G$ and $(M \times I-Q_1)/G$ in $(M \times I)/G$. This condition works for at least some higher-dimensional hot-dog problems.

REFERENCES

[A] Alexander, J.C., Bifurcation of zeroes of parameterized functions, Jour. of Func. Anal., to be published.

[A-F] Alexander, J.C. and Fitzpatrick, P.M., Bifurcation of fixed points of condensing operators, to be published.

[A-Y] Alexander, J.C. and Yorke, J.A., The homotopy continuation method: numerically implementable topological procedures, Trans. AMS, to be published.

[C-MP-Y] Chow, S.N., Mallet-Paret, J. and Yorke, J.A., Finding zeroes of maps: homotpoy methods that are constructive with probability one, to be published.

[K-L-Y] Kellogg,R.B., Li, T.Y. and Yorke, J.A., A method of continuation for calculating a Brouwer fixed point, in *Fixed Points, Algorithms and Applications* , S. Karamadian (ed.), Academic Press, New York, 1977.

Kellogg,R.B., Li, T.Y. and Yorke, J.A., A constructive proof of the Brouwer Fixed Point Theorem and computational results, SIAM Jour. Num. Anal. 13 (1976), 473-483.

[S] Smale, S., Convergent processes of price adjustment and global Newton methods, Jour. Math. Econ. 3 (1976), 1-14.

A HOMOTOPY-METHOD FOR THE CALCULATION OF ALL ZEROS OF ZERO-DIMENSIONAL POLYNOMIAL IDEALS

F. J. Drexler

Institute for Mathematics
University of Linz/Munich
Munich, Federal Republic of Germany

0. INTRODUCTION

0.1 This paper is concerned with a homotopy-method (continuation method) for calculating the complete set of zeros of zero-dimensional polynomial ideals, i.e. with computing all zeros of systems consisting of n polynomial equations in n unknowns on the only premises that the set of zeros is discrete. Higherdimensional sets of zeros are not covered by this method. The coefficients of the polynomials are in general complex numbers. This guarantees a maximum of universality for the application.

ISBN 0-12-729250-0

0.2 To prove the necessary theorems one has to make use of the algebraic, geometrical or functional-analytic point of view of the problem, i.e. the polynomials of the system are considered as a basis of a polynomial ideal in the ring of complex numbers, or as representations of (n-1)-dimensional surfaces in a n-dimensional affine space where the zeros of the system are the coordinates of the points of intersection. Finally the polynomial system can be considered as a nonlinear operator mapping the n-dimensional complex space C^n into itself.

0.3 A fundamental advantage of the method described here is that bifurcation can be avoided completely, so that extensive numerical computation is not necessary. Bifurcation frequently appears with the solution of nonlinear problems with the aid of continuation methods. To continue the computation in such a case one has to solve the bifurcation-equation which often leads to a system of polynomial equations. Such systems can be solved by the method described here.

0.4 First I will give a precise formulation of the problem, some notations and a short description of the method. Then I will describe a procedure which makes it possible to reduce statements about the solutions of a system of n polynomial equations in n unknowns to statements about the zeroes of one polynomial in one unknown. This procedure is purely theoretical and not at all suited for the practical computation on a computer, but it makes the proof of the practicability of the continuation method very easy.

0.5 First of all one has to show that bifurcation can be avoided for every given problem. This will be one of the main points of this paper. Then one has to show that other problems like poles or incontinuous solutions do not occur. Finally the numerical side of the method will be touched where I will show for example that using NEWTON's algorithm one will get every solution of an arbitrary polynomial system (of the above described type) in a finite series of steps. Besides that I will give some numerical results and discuss some problems occurring with the numerical treatment of the continuation method.

1. THE PROBLEM AND A METHOD FOR ITS SOLUTION

1.1 Let

1.1.1 $Tx = o,\ x = (x_1,\ldots,x_n)^T$

be a system of n polynomial equations in the n unknowns $x_1,\ldots,x_n$ having a discrete set of solutions. This implies that this set is finite.

One can consider T as a nonlinear operator mapping C^n into itself. T then is continuous and arbitrarily Frechet-differentiable.

1.2 The system Tx always can be written in such a way that it can be decomposed in the following manner:

1.2.1 $Tx = \Phi x + \Psi x$

with

1.2.2 $$\Phi x = \begin{pmatrix} f_1(x_1) \\ \vdots \\ f_n(x_n) \end{pmatrix} \qquad \Psi x = \begin{pmatrix} g_1(x_1,\ldots,x_n) \\ \vdots \\ g_n(x_1,\ldots,x_n) \end{pmatrix}$$

$f_k(x_k)$ and $g_k(x_1,\ldots,x_n)$ are for every $k = 1,\ldots,n$ polynomials in one respectively n unknowns for which

1.2.3 $\text{degree}(f_k) \geq \text{degree}(g_k)$, $k = 1,\ldots,n$

holds. Moreover f_k shall have only simple zeroes.

1.3 If Tx doesn't have the decomposition (1.2.1, 1.2.2, 1.2.3) anyhow one can attain it by adding proper polynomials to Φx and subtracting the same polynomials from Ψx not altering the set of solutions of $Tx = o$.

1.4 To get all solutions of the system $Tx = \Phi x + \Psi x = o$ it is embedded in a set of systems in the following way:

1.4.1 $T(x,t) = \Phi x + t\Psi x = o$

The system $T(x,0) = \Phi x = o$ can as one easily realizes be solved by conventional methods. To obtain all solutions one has to compute the zeros of the single polynomials $f_k(x_k)$

using a suitable method. Let l_{kj_k}, $k = 1,\dots,n$, $j_k = 1,\dots,$ degree(f_k) be the zeros of the polynomial $f_k(x_k)$. Then all vectors

1.4.2 $(l_{1j_1},\dots,l_{nj_n})^T$

are solutions of $\Phi x = o$, and aside from them there are no other solutions. Their number then equals the product of the degrees of the polynomials f_k. Because of the l_{kj_k} being simple zeros all vectors (1.4.2) are different by pairs and hence simple solutions of $T(x,0) = o$.

The system $T(x,1) = o$ is equivalent to the system $Tx = o$.

1.5 Let $l_j(t)$, $j = 1,\dots, N$ be the solutions of the system (1.4.1) for all $t \in W$ where W is a curve in the plane of complex numbers connecting the points $0 + i0$ and $1 + i0$ continuously. This curve is called "Einbettungsweg" (curve of embedding). Hence the following holds:

1.5.1 $T(l_j(t),t) = o$ for all $t \in W - \{1\}$ and $j = 1,\dots,N$

The point $1 \in W$ has to be discussed separately.

1.6 One can choose $W - \{1\}$ in such a way that :

1.6.1 For all $t \in W - \{1\}$ no solution $l_j(t)$ has a pole, i.e. $\| l_j(t) \| < \infty$ for all $j = 1,\dots,N$.

1.6.2 For all $t \in W - \{1\}$ the solutions $l_j(t)$ are different by pairs and hence definite.

1.6.3 For all $t \in W - \{1\}$ the $l_j(t)$ are continuous and arbitrarily often differentiable functions of the complex variable t.

The conditions need not be fulfilled for t = 1, i.e. there may be poles and multiple solutions.

1.7 The condition (1.6.2) can be replaced by the claim that for all $t \in W - \{1\}$ the functional determinant det $T'_x(l_j(t),t)$ is not equal zero because this implies by the implicite function theorem that the solutions $l_j(t)$ are definite.

1.8 The proof that there is always a curve W fulfilling the conditions (1.6.1, 1.6.2, 1.6.3) is carried out by showing that the number of all $t \in C$ for which they are not fulfilled is finite.

1.9 To come to the solutions of Tx = o one has to determine numerically the single branches of solutions $l_j(t)$, $t \in W$. For this purpose one computes iteratively for a number of successive values $t_k \in W$, $k = 0,\ldots,q$, $t_0 = 0$, $t_q = 1$ the solutions $l_j(t_k)$ using the values $l_j(t_{k-1})$ as a first approximation for the iteration method. The procedure respectively starts at the known initial solutions $l_j(t_0) = l_j(0)$ (1.4.2). As iteration method only NEWTON's algorithm is considered here.

1.10 The conditions (1.6.1, 1.6.2, 1.6.3) make sure that the above described method for the determination of the solutions of Tx = o is feasible. The following part of the paper shows that they are fulfilled for all polynomial systems of type (1.4.1). Because the initial problem has at least as many solutions as Tx = o it is guaranteed that one gets every solution of Tx = o. If the initial problem T(x,O) = o has more solutions than Tx = o condition (1.6.1) or (1.6.2) is not fulfilled for t = 1 any more. In the first case there are solution-functions $l_j(t)$ which have poles there and hence lead to no solutions of Tx = o. In the second case Tx = o has multiple solutions which may be computed, however, using NEWTON's method.

2. THEORY OF THE ALGORITHM

2.1 In this part I will describe a method which makes it possible to reduce statements about the solutions of systems of polynomials in n unknowns to statements about the zeros of polynomials in one unknown. For the theory of this method see [5], [6], [9], [13].

2.2.1 Definition: A polynomial

2.2.2 $$R_k(x_k,t) = \sum_{j=0}^{N} K_j^{(k)}(t)\, x_k^j$$

where the $K_j^{(k)}(t)$, $j = 0,\ldots,N$ are polynomials in the variable t with coefficients from C is called "Resultante" of the system $T(x,t)$ with respect to the variable x_k if the following is true:

To every zero $l_j(t)$ of $T(x,t)$ there corresponds exactly one zero $a(t)$ of $R_k(x_k,t)$ in such a manner that $a(t)$ is the k-th component of $l_j(t)$. On the other hand to every zero $a(t)$ of $R_k(x_k,t)$ there corresponds at least one zero $l_j(t)$ of $T(x,t)$ in the same way.

2.2.3 Note: The set of zeros of $R_k(x_k,t)$ is the same as the set of the k-th components of the solutions of $T(x,t) = o$. It isn't excluded, however, that there are more solutions of $T(x,t) = o$ than of $R_k(x_k,t)$ because several zeros of $T(x,t)$ may have the same k-th component.

2.3 The very extensive proof of the existence of the Resultante is contained in [5], pp. 44 -50. In this place I only will sketch it with respect to our problem. One can look at the polynomials $h_1(x_1,\ldots,x_n,t),\ldots,h_n(x_1,\ldots,x_n,t)$ of the system $T(x,t)$ as a basis of an ideal

2.3.1 $$\mathfrak{a} = (h_1,h_2,\ldots,h_n)$$

in the ring of polynomials $K[x_1,\ldots,x_n]$ where K is the field of quotients $C[t]/C[t]$, because the coefficients of the polynomials $h_1,\ldots,h_n$ are polynomials in t themselves. The zeros of $T(x,t)$ are then the same as the zeros of the ideal $\mathfrak{a}$. Starting with $\mathfrak{a} = \mathfrak{a}_0$ one can form a series of elimination-ideals $(\mathfrak{a}_j)$, $j = 1,\ldots,n-1$ where every following ideal contains one variable less then the preceding. The last elimination-ideal in this series, namely $\mathfrak{a}_{n-1}$, is then generated by one polynomial in only one unknown and is equivalent to $R_k(x_k,t)$. For this series of elimination-ideals the following theorem is valid (see [5]):

2.3.2 Theorem: To every zero $(a_1,\ldots,a_k,\ldots,a_n)$ of $\mathfrak{a}$ there corresponds exactly one zero a_k of $\mathfrak{a}_{n-1} = \mathfrak{a} \cap K[x_k]$ and to every zero of $\mathfrak{a}_{n-1}$ there corresponds at least one zero of $\mathfrak{a}$ in the same way.

2.3.3 The geometrical meaning of this algebraic elimination-method is that the zeros of $T(x,t)$ for every given t are projected on the subspace $x_1 = x_2 = \ldots = x_{k-1} = x_{k+1} = \ldots$ $\ldots = x_n = 0$ of C^n.

2.4 For the Resultante then holds:

2.4.1 $R_k(l_{jk}(t),t) = 0$ for all $t \in W - \{1\}$ and all $k = 1,\ldots,n$

for every solution-function

2.4.2 $l_j(t) = (l_{j1}(t),\ldots,l_{jn}(t))^T$

of $T(x,t) = o$.

2.5 Finally let me point at another algebraic notion which I will need in the following, namely that of the SYLVESTER-determinant of two polynomials in the same unknown. This is a determinant formed by the coefficients of the two polynomials whose disappearance is necessary and sufficient for the existence of joint zeros of the two polynomials.

2.6 To illustrate the preceding part let me give an example for the formation of the Resultante:

2.6.1 Example: Let

$$T(x,t) = \begin{pmatrix} h_1 \\ \\ h_2 \end{pmatrix} = \begin{pmatrix} x_1^2 + 1 + tx_2^2 \\ \\ x_2^2 - 1 - tx_1^2 \end{pmatrix} = o$$

Then

$$R_1(x_1,t) = h_1 - th_2 = (1 + t^2)x_1^2 + 1 + t$$

$$R_2(x_2,t) = th_1 + h_2 = (1 + t^2)x_2^2 - 1 + t$$

The set of zeros of R_1 resp. R_2, i.e. the set of the first resp. second components of the solutions of $T(x,t) = o$ is then

$$\{ia, -ia\} \quad \text{resp.} \quad \{b, -b\}$$

with

$$a = \left(\frac{1 + t}{1 + t^2}\right)^{1/2} \qquad b = \left(\frac{1 - t}{1 + t^2}\right)^{1/2}$$

As solution of $T(x,t) = o$ one gets

$l_1(t) = (ia, b)$, $\quad l_2(t) = (-ia, b)$

$l_3(t) = (ia, -b)$, $\quad l_4(t) = (-ia, -b)$

2.7 In the following section I will show, that the set of all $t \in C$ for which poles of solution-functions may appear is finite.

2.7.1 Lemma: Let $l_{jk}(t)$, $j = 1,\ldots,N$ be the set of zeros of the Resultante $R_k(x_k,t)$. Then for a given $t_p \in C$

$$\lim_{t \to t_p} |K_N^{(k)}(t)| = 0$$

is valid if there is at least one $l_{j_o k}(t) \in \{l_{jk}(t)\}$ with

$$\lim_{t \to t_p} |l_{j_o k}(t)| \to \infty$$

and vice versa.

2.7.2 In other words poles of solution-functions appear exactly at that points where the leading coefficients of (2.2.2) vanish for at least one $k \in \{1,\ldots,n\}$.

2.7.3 To prove that lemma one has to decompose the Resultante (2.2.2) in a product of linear factors. Because the $K_N^{(k)}(t)$, $k = 1,\ldots,n$ are polynomials in t, their number of zeros is finite, i.e. the following theorem holds

2.7.4 Theorem: The number of all $t \in C$ for which poles of solutions $l_j(t)$ appear is finite for all $j = 1,\ldots,N$.

2.7.5 Note: For the rest we think of W as chosen in such a way that no poles appear on it.

2.8 We now come to continuity and differentiability of the solutions of $T(x,t) = o$.

2.8.1 Theorem: The single components $l_{jk}(t)$, $k = 1,\ldots,n$ of the solution-functions $l_j(t)$, $j = 1,\ldots,N$ are continuous for all $t_s \in C$ for which none of the solution-functions has a pole.

2.8.2 Proof: To prove that theorem one has to keep in mind that the zeros of polynomials are continuous functions of the coefficients. Because the coefficients are continuous in t themselves it follows that the zeros $l_j(t)$ are continuous functions of the variable t.

2.9.1 Theorem: The single components $l_{jk}(t)$, $k = 1,\ldots,n$ of the solution functions $l_j(t)$, $j = 1,\ldots,N$ are arbitrarily often continuously differentiable for all $t \in C$ for which they have no pole and for which they are simple solutions of $R_k(x_k,t)$.

For the derivative of $l_{sk}(t)$, $s \in \{1,\ldots,N\}$ with respect to t the following holds

2.9.2 $$l'_{sk}(t) = \frac{\sum_{j=0}^{N} K_j^{(k)}(t)\, l_{sk}(t)^j}{-K_N^{(k)}(t) \prod_{\substack{j=1 \\ j \neq s}}^{N} (l_{sk}(t) - l_{jk}(t))}$$

$$s = 1,\ldots,N,\ l_{sk}(t) \neq l_{jk}(t) \text{ for all } j = 1,\ldots,N,\ j \neq s$$

2.9.3 Proof: If one differentiates both sides of the identity

$$\sum_{j=0}^{N} K_j^{(k)}(t)\, x_k^j = K_N^{(k)}(t) \prod_{j=1}^{N} (x_k - l_{jk}(t))$$

with respect to t and then takes $x_k = l_{sk}(t)$ one gets (2.9.2) by separation with respect to $l'_{sk}(t)$.

2.9.4 Note: One may check formula (2.9.2) using example (2.6.1) where the functions $l_{jk}(t)$ are explicitely known. Remark further that $l'_{sk}(t)$ is continuous and arbitrarily often differentiable simply by observing the rules for differentiating a quotient of continuous functions.

2.10.1 Definition: A function $l_{jk}(t)$ is called an identically multiple zero of $R_k(x_k,t)$ if it is for all $t \in C$ multiple zero of $R_k(x_k,t)$.

2.10.2 Note: About the differentiability of identically multiple zeros theorem (2.9.1) makes no statement though such functions may be differentiable as examples show.

2.11.1 Lemma: Two functions $l_{j_1k}(t)$ and $l_{j_2k}(t)$ $j_1, j_2 = 1,\ldots,N$ have the same value in at most

2.11.2 $\quad (2N - 1) \max_{p} (\text{degree } (K_p^{(k)}(t)))$

points $t \in C$ if there are no identically multiple solutions of $R_k(x_k, t) = 0$.

2.11.3 Proof: One proves this lemma using the fact that multiple zeros of $R_k(x_k, t)$ are at the same time zeros of the derivative $R'_{k_x}(x_k, t)$, i.e. the SYLVESTER-determinant of the polynomials R_k and R'_{k_x} vanishes. This determinant is a polynomial in t which has at most a finite number of zeros if it isn't identically zero. (2.11.2) is an upper bound of the number of zeros.

2.12 To illustrate the preceding section I will give a short example:

2.12.1 Example: Let

$$R_k(x_k, t) = (t^2+1)x_k^2 + x_k - t^2$$

which has the zeros

$$l_{1k}(t) = \frac{t^2}{1 + t^2} \quad \text{and} \quad l_{2k}(t) = -1$$

with

$$R'_{k_x}(x_k, t) = 2(t^2+1)x_k + 1$$

one gets the SYLVESTER-determinant

$$\begin{vmatrix} t^2+1 & 1 & -t^2 \\ 0 & 2(t^2+1) & 1 \\ 2(t^2+1) & 1 & 0 \end{vmatrix} = (t^2+1)(1+4t^2(t^2+1))$$

Two of the four zeros, namely $t = i$ and $t = -i$ are points of poles (leading coefficient of R_k disappears) and are not interesting here. The other zeros are $t = \frac{1}{2}\sqrt{2}i$ and $t = -\frac{1}{2}\sqrt{2}i$ and exactly at that points $l_{1k}(t)$ and $l_{2k}(t)$ have the same value.

2.13 Using theorem (2.9.1) and lemma (2.11.1) we get

2.13.1 Theorem: All solution-functions are, with the exception of finitely many points, arbitrarily often differentiable if there exists a $t_u \in C$ with $l_{pk}(t_u) \neq l_{qk}(t_u)$ for all $p,q = 1,\dots,N$, $p \neq q$.

2.14.1 Lemma: For all systems of polynomial equations (1.4.1) there exists a nonsingular, linear, homogeneous transformation $A \in C^{nxn}$ so that the k-th components of the solutions of

2.14.2 $\quad T(\tilde{x},0) = \Phi\tilde{x} = o$, $\tilde{x} = Ax$

are different by pairs for all $k = 1,\dots,n$.

2.14.3 Proof: We prove this theorem by specifying the transformation $A = (a_{ik})$.

Because of the construction of the method all solution-vectors $l_j(0)$, $j = 1,\dots,N$ of $\Phi x = o$ are different by pairs.

As solutions of $\Phi\tilde{x} = o$ one gets $\tilde{l}_j(0) = Al_j(0)$.

Let $d_{pq} := l_p(0) - l_q(0)$ for all $p,q = 1,\dots,N$, $p \neq q$ and let a_s be a vector whose s-th component is notequal zero for some $s \in \{1,\dots,n\}$ with $d_{pq} \cdot a_s = c_{pqs}$, $|c_{pqs}| \geq c > 0$ for all $p,q = 1,\dots,N$. $p \neq q$. (Here $u \cdot v$ means the sum of products of the corresponding components of the vectors u and v.)

Finally let $\varepsilon_j = \dfrac{c-b_j}{\max\limits_{p,q}|d_{pq,j}|}$, $c > b_j > 0$ for $j \neq s$

and $\varepsilon_s = 0$. ($d_{pq,j}$ means the j-th component of d_{pq}.)
If the denominator equals zero one may choose $\varepsilon_j \neq 0$ arbitrarily.

The matrix A whose rows are the vectors $a_j := a_s + e_j$ $j = 1,\dots,n$ then has the required properties. Here e_j is defined as $e_j := (0,\dots,0, \varepsilon_j, 0,\dots,0)^T$. It is easy to verify that using this transformation one has $\tilde{l}_{pk}(0) \neq \tilde{l}_{qk}(0)$ for all $p,q = 1,\dots,N$, $p \neq q$, $k = 1,\dots,n$.

Hence we get immediately

2.15.1 Theorem: The solution-functions $l_j(t)$, $j = 1,\dots,N$ of every polynomial system $T(x,t) = o$ are with finitely many exceptions arbitrarily often continuously differentiable. This holds especially for $t = 0$.

2.15.2 Proof: Theorem (2.13.1) applies with $t_u = 0$ and $\tilde{l}_j(t)$ and by $l_j(t) = A^{-1}\tilde{l}_j(t)$ it is easy to see that $l_j(t)$ has the same differentiability-properties as $\tilde{l}_j(t)$.

2.16 Finally we have to show that the JACOBI-matrix

2.16.1 $T'_x(l_j(t),t)$ $j = 1,\dots,N$

is nonsingular with the exception of at most finitely many points. Denotes

2.16.2 $D(x,t) := \det (T'_x(x,t))$

2.16.3 $D_j(t) := D(l_j(t),t)$ for all $j = 1,\dots,N$

and let $G_j \subset C$ be a connected region with $0 \in G_j$.

Then

2.17.1 Lemma: $D_j(0) \neq 0$ for all $j = 1,\dots,N$.

2.17.2 Proof: By construction of our method

$$D_j(0) = D(l_j(0),0) = \det (T'_x(l_j(0),0)) =$$

$$= \prod_{k=1}^{n} f'_k(l_{jk}(0)) \neq 0 \text{ for all } j = 1,\dots,N \text{ because the}$$

polynomials $f_k(x_k)$ possess simple zeros only.

2.18 Now we are able to prove the following

2.18.1 Theorem: The JACOBI-matrix $T'_x(l_j(t),t)$ has only finitely many points of singularity for all $j = 1,\dots,N$.

2.18.2 Proof: It is sufficient to show that $D_j(t)$ has only finitely many zeros in a region $G_j \subset C$.

$D(x,t)$ is a polynomial in the variables $x_1, \ldots, x_n, t$. Hence $D_j(t)$ is a sum of products of the single components of $l_j(t)$ among one another and with polynomials in the variable t. Since these components are differentiable with only the exception of finitely many points the same holds for the complex function $D_j(t)$. Hence $D_j(t)$ is a regular function in a connected region $G_j \subset C$ and by that, using a well-known theorem of complex analysis, $D_j(t)$ has in G_j at most finitely many zeros pretending it isn't identically zero. By lemma (2.17.1) this can not be.

2.18.3 Note: One may choose G_j as set of all $t \in C$ with $l_j(t)$ differentiable and $|t| \leq R$, R sufficiently large.

2.19 Summarizing the preceding chapter we can state that one may choose W always in such a manner that conditions (1.6.1, 1.6.2, 1.6.3) are fulfilled.

3. IMPLEMENTATION OF THE ALGORITHM

3.1 In this section I will show that one may follow every branch of solution from $t = 0$ to $t = 1$ along a path W in the complex plane using NEWTON's method and needing only a finite number of steps. We pretend that W fulfills conditions (1.6.1, 1.6.2, 1.6.3). Branches of solutions not fulfilling (1.6.1) for $t = 1$ need not be considered further and for the other the following consideration is valid for all $t \varepsilon W$ with $|1 - t| > \varepsilon$.

For all solutions then holds

3.2.1 $\|T'_x(l_j(t),t)^{-1}\| \le b < \infty$, $b \varepsilon R$, $t \varepsilon W$

3.2.2 $\|\Psi(l_j(t))\| \le c < \infty$, $c \varepsilon R$, $t \varepsilon W$

3.2.3 $\|\Psi'_x(l_j(t))\| \le d < \infty$, $d \varepsilon R$, $t \varepsilon W$

3.2.4 $\|T'_x(x_1,t) - T'_x(x_2,t)\| \le L\|x_1 - x_2\|$

$L < \infty$, for all x_1, $x_2 \varepsilon G \subset C^n$, G closed and for all $t \varepsilon C$ with $|t| \le r_0$, $r_0 \varepsilon R$ arbitrary.

The last estimation holds because the operator T is arbitrarily often Frechet-differentiable.

3.3 With the estimations above one can show that using NEWTON's method one can compute the solution $l_j(t_k)$ of $T(x,t_k) = 0$, $t_k \varepsilon W$ iteratively using the solution $l_j(t_{k-1})$ of $T(x,t_{k-1})$, $t_{k-1} \varepsilon W$ as a first approximation where the step-width

3.3.1 $s := t_k - t_{k-1}$

always can be chosen fulfilling

3.3.2 $|s| \ge s_0 > 0$ for all $t_{k-1} \varepsilon W$.

3.4 To show this one needs the theorem about the convergence of NEWTON's method from [7], pp. 572. One obtains s_o as solution of the equation

3.4.1 $$(1 + \frac{1}{1 - bds_o})^2 s_o = \frac{1}{2b^2Lc}$$

fulfilling

3.4.2 $$0 < s_o < \frac{1}{bd}$$

3.5 According to [7] the following statement holds globally: For every $t_k \in W$ with $|t_k - t_{k-1}| \leq s_o$ there exists a unique solution $l_j(t_k)$ of $T(x,t_k) = o$ to which NEWTON's algorithm converges using $l_j(t_{k-1})$ as a first approximation.

3.5.1 For practical computation the bound s_o for the stepwidth proves to be too small in most cases.

3.6 Finally we have to treat the till now excluded case where

3.6.1 $$\det (T'_x(l_j(1),1)) = 0$$

for some solution $l_j(t)$.

In this case one can compute $l_j(t_{k-1})$ for all $t_{k-1} \in W$ with $|t_{k-1} - 1| \geq \varepsilon$, especially for some

3.6.2 $$t_{k-1} \in \{t \in C \mid |t - 1| = \varepsilon, \varepsilon > 0\}$$

Starting with the value $l_j(t_{k-1})$ one can compute $l_j(1)$ using

NEWTON's method though (3.6.1) holds (see [14]). One only has to choose ε sufficiently small. Convergence in this case is linear.

3.7 This completes the proof of the practicability of the method described in this paper for solving polynomial systems of type (1.1.1).

4. NUMERICAL RESULTS

4.1 In this section I will give some numerical results. Based on the considerations of the preceding sections I have made an ALGOL-program and tested it with about 200 examples. The number of unknowns lay between 2 and 8, the number of solutions totaled to 200 in eight-dimensional problems. In all tested problems all solutions were found though sometimes the step-width had to be chosen very small ($|s| \leq 10^{-3}$).

4.2 In the following I shall discuss in some detail the numerical difficulties which have occured.

4.2.1 The program is designed in such a way as to choose as the Einbettungsweg W the interval [0,1] firstly and only if there come up difficulties W is shifted into the complex plane. Moreover it is tried to solve the problem with a fixed step-width s with $|s| \geq 0.1$ and to reduce it only when necessary. The choice of the Einbettungsweg in the complex plane and the reduction of the step-width is performed by the program itself.

4.2.2 Numerically without problems in most cases are systems with complex coefficients and systems with a linear operator Ψx. In general computation is successful with relatively large step-widths. Numerically much more difficult are problems where $T(x,0) = o$ has more solutions than $T(x,1) = Tx = o$. As I have shown there are poles of solutions at $t = 1$. This case occurs rather frequently when the system is not of type (1.2.1) originally but has to be transformed (see 1.3). For t in the neighbourhood of 1 there are very small step-widths necessary.

4.3 In an overwhelming quantity of the tested examples with real or complex coefficients there were only some real solutions compared with much more complex solutions especially with a greater number of unknowns.

4.4 Some problems occured with the choice of the Einbettungsweg in the complex plane. The program procedes herein as follows: A place $t_q \in C$, $t_q \neq 1$ is passed in some distance if at t_q one of the following cases happens:

4.4.1 The difference of two succeeding values in NEWTON's iteration is still too large after a given number of iteration-steps.

4.4.2 The determinant of the JACOBI-matrix T'_x exceeds the value of 10^9.

4.4.3 The determinant of the JACOBI-matrix increases from one iteration-step to the next more than fifefold.

4.4.4 The determinant of the JACOBI-matrix decreases from one iteration-step to the next to half of its value and increases in the following step to its original value or still more.

4.4.5 The imaginary part of a component varies for some fixed t by a factor 10^{-4} in the iteration on this step whereas the real part doesn't alter very much.

4.5 The determination of W eventually in combination with a reduction of the step-width on grounds of the above described criterions (4.4.1 - 4.4.5) proves to be sufficient to get all solution of a polynomial system using the method of this paper.

5.1 Finally let me refer to [3] where a great number of examples may be found to this section but also to the preceding sections.

REFERENCES

[1] Brown, K. M., Gearhart, W. B., Deflation techniques for the calculation of further solutions of a nonlinear system, Numerische Mathematik 16, 334-342 (1970/71)

[2] Collatz, L., Funktionalanalysis und Numerische Mathematik, Springer-Verlag, Berlin, Heidelberg, New York (1968)

[3] Drexler, F.-J., Eine Methode zur Berechnung sämtlicher Lösungen von Polynomgleichungssystemen, Dissertation, Linz, München (1976)

[4] Feilmeier, M., Lory, P., Scheuring, H., Wacker, H. J., DFG-Berichte 1, 2, 3, "Einbettung", (1973/74/75)

[5] Gröbner, W., Moderne algebraische Geometrie, Die idealtheoretischen Grundlagen, Springer-Verlag, Wien, Innsbruck (1949)

[6] Gröbner, W., Algebraische Geometrie II, Bibliografisches Institut, Mannheim (1970)

[7] Kantorowitsch, L. W., Akilow, G. P., Funktionalanalysis in normierten Räumen, Akademie-Verlag, Berlin (1964)

[8] Krasnosel'skii, M. A., Approximate solution of operator equations, Wolters-Noordoff Publishing, Groningen (1972)

[9] Lugowski, H., Weinert, H. J., Grundzüge der Algebra, Pfalz-Verlag, Basel (1967)

[10] Sather, D., Branching of solutions of an equation in Hilbert Space, Arch. Rat. Mech. Anal. 36, 47-64 (1970)

[11] Tutschke, W., Grundlagen der Funktionentheorie, Vieweg-Verlag, Braunschweig (1969)

[12] Wacker, H. J., Nichtlineare Homotopien zur Konstruktion von Startlösungen für Iterationsverfahren (Numerische Lösung nichtlinearer partieller Differential- und Integrodifferentialgleichungen), Lecture Notes 267, 47-64, Springer-Verlag (1972)

[13] Wainberg, M. M., Trenogin, W. A., Theorie der Lösungsverzweigung bei nichtlinearen Gleichungen, Akademie-Verlag, Berlin (1973)

[14] Young, D. M., Iterative solution of large linear systems, Academic Press, New York, London (1971)

SOLUTION OF OPTIMIZATION PROBLEMS WITH NONLINEAR RESTRICTIONS VIA CONTINUATION METHODS

J. Hackl

Institut für Mathematik
Universität Linz
Linz, Austria

I. INTRODUCTION

The aim of the present paper is to give a new algorithm treating nonlinear programming problems in an infinite dimensional space. The algorithm should be useful in nonlinear programming (large scale) and control theory. By a *Mathematical Programming Problem* we mean a problem of the following type: Given Hilbert spaces E, F_1, F_2, a closed convex set $Y \subset F_1$ and mappings $f: E \to \mathbb{R}$, $g: E \to F_1$, $h: E \to F_2$. Find $x \in E$ which minimizes f on the set $Z := \{x \in E \mid g(x) \in Y \wedge h(x) = o\}$

ISBN 0-12-729250-0

For mathematical programming problems we also write

$$(P) := \left\{ \begin{array}{l} f(x) = \text{Min !} \\ g(x) \in Y \\ h(x) = o \end{array} \right\}$$

In this paper we do not restrict ourselves to F_1, F_2 finite dimensional, as in many control problems an infinite number of restrictions occurs. But we want f,g,h to be differentiable in a strong sense and g,h to be compact. The last assumption is fulfilled in many control problems too, e.g. if restrictions in state-space variables are treated in control space. Although some algorithms are restricted to equality or inequality constraints only or linear constraints there is a large number of methods treating mathematical programming problems (in most cases E, F_1, F_2 are assumed to be finite dimensional, Y positive or negative octant in F_1). Most important are gradient projection methods (see e.g. [1], [2], [3]), penalty and barrier-function and lagrangian methods (see e.g. [4], [5], [6], [7], [8], [9], [1o]). As a curiosum see [19] .

In this paper we are interested in a solution of problem (P) in a certain sense. x_o is accepted as a solution if x_o is nearly feasible and nearly stationary. Both ideas will be formulated exactly in this paper. We are not interested to give necessary or sufficient conditions for optimality.

II. BASIC CONCEPT AND THEOREMS

Most optimization technics work in the following way. Starting in a feasible point we look for a feasible and descending (if we want to minimize) direction. If such a direction does not exist, the algorithm terminates in a stationary point; otherwise the direction defines a ray, where a new point is chosen. A not constructive result about such directions is proved in [11] :

Theorem 1: Let E, F_1, F_2 be real Banachspaces. $Y \subset F_1$ convex $\wedge$ $\wedge$ int $Y \neq \emptyset \wedge g: E \to F_1$ Hadamarddifferentiable in $x_o \varepsilon E \wedge$ $\wedge h: E \to F_2$ Gateauxdifferentiable in open domain D, $x_o \varepsilon D \wedge$ $\wedge g(x_o) \varepsilon Y \wedge h(x_o) = o \wedge h_x(x_o)$ surjective $\wedge$

$$\wedge \bigvee_{\Delta x \varepsilon E} g(x_o)+g_x(x_o)\Delta x \varepsilon \text{ int } Y \wedge h(x_o)+h_x(x_o)\Delta x = o \implies$$

$$\implies \bigvee_{\substack{r:[o,\tau[\to E \\ \tau > o}} \bigwedge_{t \varepsilon [o,\tau[} g(x_o+t\Delta x+r(t)) \varepsilon Y,\ h(x_o+t\Delta x+r(t))=o$$

$$\lim_{t \to +o} r(t)/t = o$$

Remark: Definitions of Hadamard-, Gateauxdifferentiable see e.g. [12] .

As in practical computation neither x_o nor x fulfills such restrictive conditions like Theorem 1, we try to give definitions and theorems more fitting practical computation.

Definition 2: Let $x \varepsilon E$, $\varepsilon \in \mathbb{R}_o^+$. x is called ε-feasible for (P)

$$:\Longleftrightarrow \quad \|g(x)^-\| \leq \varepsilon \wedge \|h(x)\| \leq \varepsilon$$

Throughout this paper y^+ means the nearest point in Y to $y \in F_1$ and $y^- := y - y^+$. See Figure 1.

If Y is the positive octant in $\mathbb{R}^n$, y^+, y^- has the usual meaning. Next example shows that existence of ε-feasible points for all $\varepsilon \in \mathbb{R}^+$ need not imply the existence of a o-feasible point.

Example: $E := \mathbb{R}^2$, $F_1 := F_2 := \mathbb{R}$, $g(x_1, x_2) := -x_2$, $Y := \mathbb{R}_o^+$
$h(x_1, x_2) := x_1^2 x_2 - 1$. See Figure 2.

Further we need differentiability conditions at every point x_o in a ball with radius $\sqrt{r}$.

A mapping $t: E \to F$ fulfills condition [D] at $x_o \in E$ with constants $C, r \in \mathbb{R}^+$ iff t is Frechetdifferentiable in x_o and

$$\bigwedge_{\Delta x \in E} \quad \|\Delta x\|^2 \leq r \implies \|t(x_o + \Delta x) - t(x_o) - t_x(x_o)\Delta x\| \leq C\|\Delta x\|^2$$

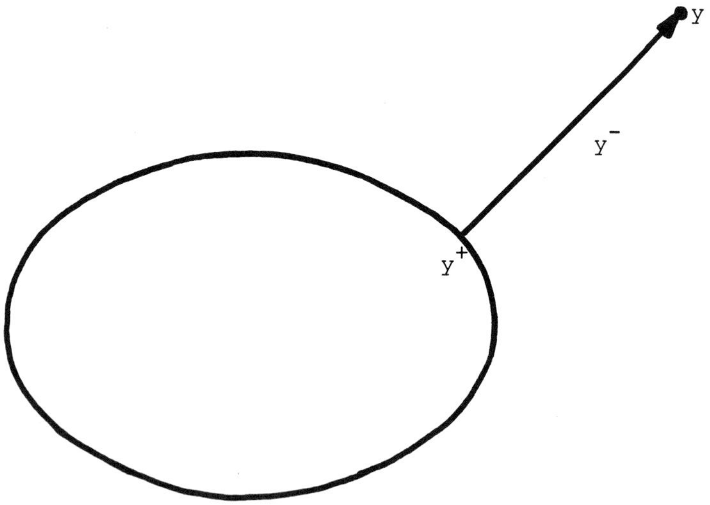

FIGURE 1.

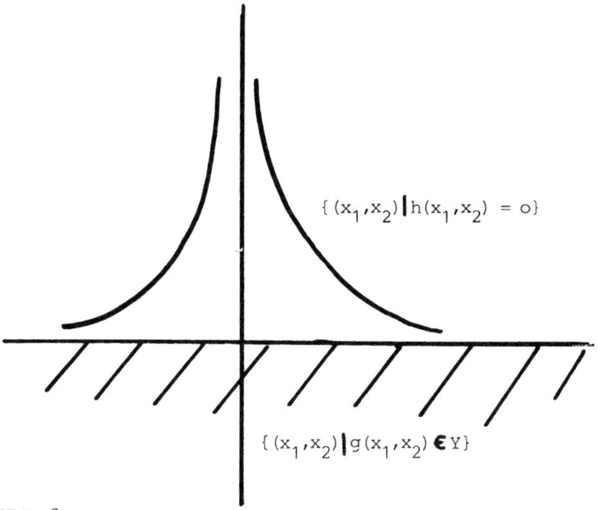

FIGURE 2.

At x_o condition [DIF] holds with constants $C, r \in \mathbb{R}^+$, iff f Frechetdifferentiable in x_o and [D] holds with constants C,r and $t \in \{g,h\}$. f,g,h Frechetdifferentiable holds throughout this paper.

Now we can linearize problem (P) at x_o.

Definition 3: Let $x_o \in E$, $r \in \mathbb{R}^+$. Then we define:

$$(LP) := \left\{ \begin{array}{l} f_x(x_o)\Delta x = \text{Min } ! \\ g(x_o) + g_x(x_o)\Delta x \in Y \\ h(x_o) + h_x(x_o)\Delta x = o \\ <\Delta x, \Delta x> \leq r \end{array} \right\}$$

Next we give a series of problems representing simplifications of problem (LP).

Definition 4: Again let $x_o \in E$, $r \in \mathbb{R}^+$. Then we define

$$(LP_o) := \left\{ \begin{array}{l} f_x(x_o)\Delta x = \text{Min } ! \\ <\Delta x, \Delta x> \leq r \end{array} \right\}$$

and by induction while (LP_{k-1}) is welldefined and has a solution,

$$\bigwedge_{k \varepsilon \mathbb{N}} \Delta x_k := \text{solution of problem } (LP_{k-1})$$

$$x_k := x_o + \Delta x_k$$

$$(LP_k) := \begin{cases} (LP_{k-1}) \\ <g(x_k)^-, g(x_o) + g_x(x_o)\Delta x - g(x_k)^+> \leq o \\ <h(x_k), h(x_o) + h_x(x_o)\Delta x> = o \end{cases}$$

The algorithm above will be called ASLP (Approximate solution of LP). Some considerations show ASLP to be strongly related to an algorithm given in [13] .

Lemma 5: If there exists an integer k, such that (LP_{k-1}) has no feasible solution, then (LP) has no feasible solution. Furthermore, if (LP) has a feasible point, for each integer k (LP_{k-1}) has a solution.

Proof: See Appendix.

Lemma 6: F Hilbertspace, Y closed convex subset of F, $a, b \in F$, a^-, a^+, b^-, b^+ defined as above $\Longrightarrow$

$$\Longrightarrow <a^-, a^+ - b^+> \leq \left\| \frac{b-a}{2} \right\|^2$$

$$\| a^- - b^- \| \leq \| a-b \|$$

$$\| a^+ - b^+ \| \leq \| a-b \|$$

Proof: See Appendix.

Now we show a first result about ASLP.

Theorem 7: Let [DIF] be valid at $x_o \varepsilon E$ with constants $C, r \in \mathbb{R}^+$ and g,h compact mappings and (LP) has a feasible point, then

1. $\bigwedge_{\varepsilon \; \mathbb{R}^+} \bigvee_{n_o \varepsilon \; \mathbb{N}} \bigwedge_{n \geq n_o} \quad x_n \quad (\varepsilon + Cr)$-feasible for (P)

2. $\bigwedge_{\varepsilon \in \mathbb{R}^+} \bigvee_{n_o \varepsilon \mathbb{N}} \bigwedge_{n \geq n_o} \Delta x_n$ $(\varepsilon+Cr)$-feasible for (LP)

with x_n, Δx_n as introduced in Definition 4.

Proof: See Appendix.

Theorem 7 tells us about the quality of the solution of ASLP if ASLP has a solution, in Theorem 9 of this paper we want to give conditions quaranteeing ASLP has solutions.

Definition 8: Let $x_o \varepsilon$ E. Now we define

$$ZLP(x_o) := \{\Delta x \varepsilon E \,|\, g(x_o)+g_x(x_o)\Delta x \varepsilon Y \wedge h(x_o)+h_x(x_o)\Delta x = o\}$$

x_o is called a *regular point* $\Longleftrightarrow ZLP(x_o) \neq \emptyset$.

Regularity number R is defined as follows:

$$R := \begin{cases} \infty & \text{iff } ZLP(x_o) = \emptyset \\ o & \text{iff } x_o \text{ o-feasible for (P)} \\ \inf\{\|\Delta x\| \,/\Delta x \,\varepsilon\, ZLP(x_o)\} / \max\{\|g(x_o)^-\|, \|h(x_o)\|\} & \text{otherwise} \end{cases}$$

Regularity defined here differs from regularity used in literature of nonlinear programming. See e.g. [14,p.165], [4, p.249], [15,p.61], [16], [17].

Obviously Definition 8 gives a weaker condition than those you can find in books and papers cited above. Verification of the conditions is difficult, whether you use regularity defined here or regularity defined in literature. There are two good reasons to use regularity defined above: First we never really need to compute an inverse as it is the case in [1], [2]; second the (in this paper) unnecessary condition "$h_x(x_o)$ surjective" ([11]) together with "h compact" (see

Theorem 7) would imply F_2 to be finite dimensional euclidian space ([18, p.284]).

Theorem 9: Let $x_o \in E$ be a regular point with regularity number R, further x_o σ-feasible for (P), $r \in \mathbb{R}^+$, $r > (\sigma R)^2$ $\Longrightarrow$ (LP) has a feasible solution.

Proof: By definition 8 x_o regular implies

$$\bigwedge_{\varepsilon > o} \quad \bigwedge_{\Delta x \in ZLP(x_o)} \quad \|\Delta x\| < (R+\varepsilon)\sigma$$

$(R\sigma)^2 < r$ completes the proof.

Theorem 1o: Let $x_o \in E$, $C, r \in \mathbb{R}^+$ and g,h compact mappings and [DIF] valid at x_o with constants C,r and x_o σ_o-feasible for (P) with regularity number R $\wedge$ $(\sigma_o R)^2 < r < \frac{\sigma_1}{C}$ (*)

$\Longrightarrow$

1. $\bigvee_{n_o \in \mathbb{N}} \quad \bigwedge_{n \geq n_o}$ x_n is σ_1-feasible for (P)
2. $\bigvee_{n_o \in \mathbb{N}} \quad \bigwedge_{n \geq n_o}$ Δx_n is σ_1-feasible for (LP)

Proof: Lemma 5 and Theorem 9 guarantee the existence of $(\Delta x_n)_{n \in \mathbb{N}}$ if $(\sigma_o R)^2 < r$. Theorem 7 and $Cr < \sigma_1$ complete the proof.

r can fulfill (*) iff $(\sigma_o R)^2 < \frac{\sigma_1}{C}$ and when $\sigma_o > o$ this is equivalent to $\sigma_1/\sigma_o > \sigma_o C R^2$. Therefore $\sigma_o C R^2 < 1$ is sufficient for "improvement of feasibility" that is $\sigma_1 < \sigma_o$. To fulfill the last inequality a homotopy-method is a suitable tool.

III. IMBEDDING AND SOLUTION OF A SYSTEM OF EQUALITIES AND INEQUALITIES

In this chapter we are interested mainly in solving

$$\left\{ \begin{array}{l} g(x) \in Y \\ h(x) = o \end{array} \right\}$$

which is problem (P) omitting optimality. In this case the objective function may be omitted in (LP) as well as in (LP_k). Any feasible point of (LP_{k-1}) may be used for Δx_k. For reasons of simplicity in computation we usually take the normminimal feasible point of (LP_{k-1}).

Definition 11: $\{(P(s) \mid s \in [o,1]\}$ is called *imbedding* of problem (P): $\Longleftrightarrow$ for all $s \in [o,1]$

$$(P(s)) := \left\{ \begin{array}{l} f(x) = \text{Min !} \\ \tilde{g}(s,x) \in Y \\ \tilde{h}(s,x) = o \end{array} \right\}$$

$\tilde{g}: [o,1] \times E \to F_1$, $\tilde{h}: [o,1] \times E \to F_2$ continuous mappings $\tilde{g}(1,.) \equiv g$ and $\tilde{h}(1,.) \equiv h$ and solution of problem P(o) is known.

Theorem 12: Let $\bar{\sigma} \in \mathbb{R}^+$. $\Gamma := \{(s,x) \in [o,1] \times E \mid \max\{\|\tilde{g}(s,x)^-\|, \|\tilde{h}(s,x)\| \leq \bar{\sigma}\}$. Assume for all $s \in [o,1]$ $\tilde{g}(s,.)$ and $\tilde{h}(s,.)$ to be compact mappings and for each pair $(s,x) \in \Gamma$ there exist real numbers $R(s,x) \in \mathbb{R}_o^+$, $C(s,x)$, $r(s,x) \in \mathbb{R}^+$ such that [DIF] is valid at x with constants C(s,x),

$r(s,x) \in \mathbb{R}^+$ and regularity number $R(s,x)$. Further

$\underline{a} := \sup\{C(s,x), r(s,x) \mid (s,x) \in \Gamma\} < \bar{a} := \inf\{\sqrt{r(s,x)}/R(s,x) \mid$

$(s,x) \in \Gamma \wedge R(s,x) > o\}$ and

$$\bigvee_{L \in \mathbb{R}^+} \bigwedge_{(s,x) \in \Gamma} \bigwedge_{\Delta s \in [o,1-s]} \begin{matrix} \|g(s,x)-g(s+\Delta s,x)\| \leq L\Delta s, \\ \|h(s,x)-h(s+\Delta s,x)\| \leq L\Delta s \end{matrix}$$

$x_o \in E$ is o-feasible for $(P(o)) \Longrightarrow$

$$\bigvee_{m \in \mathbb{N}} \bigvee_{\substack{o \leq s_1 \ldots \leq s_m = 1 \\ x^1, \ldots, x^m \in E \\ \sigma_1, \ldots, \sigma_m = \bar{\sigma} \in \mathbb{R}}} \bigwedge_{1 \leq i \leq m} \begin{matrix} x^i \text{ is } \sigma_i\text{-feasible for } (P(s_i)) \\ \wedge\, x^i - x^{i-1} \text{ is solution of } (LP_{k_i}) \\ \text{at } (s_i, x^{i-1}) \text{ with suitable} \\ k_i \in \mathbb{N} \text{ and } r_i \in \mathbb{R}^+ \end{matrix}$$

Proof: See Appendix.

Theorem 12 gives sufficient conditions (strong conditions) similar to those you find e.g. in [2o] . More details about practical computation you can find in chapter 5.

IV. SOLUTION OF THE OPTIMIZATION PROBLEM

In this chapter"nearly stationary", "nearly Kuhn-Tucker-point" will be defined and criterions to find such points as well as relations to classical ideas will be given.

Definition 13: Let assumptions of Definition 4 be valid.

Let $(LP_o^o) := (LP_o)$. x_n defined like x_n in Definition 4.

$$(LP_n^o) := \begin{cases} (LP_{n+1}^o) \\ \langle g(x_n)^-, g_x(x_o)\Delta x\rangle \leq o \\ \langle h(x_n), h_x(x_o)\Delta x\rangle = o \end{cases}$$

for all $n \in \mathbb{N}$.

It is obvious that (LP_n^o) has a feasible point.

Definition 14: Let $x_o \varepsilon E$, $r \varepsilon \mathbb{R}^+$, $\alpha \varepsilon \mathbb{R}_o^+$. x_o is called (α,r)-stationary: $\Longleftrightarrow \bigvee_{n \varepsilon \mathbb{N}} \Delta x$ solution of $(LP_n^o) \wedge$ $\wedge f_x(x_o)\Delta x \geq -\alpha\sqrt{r}$.

Definition 15: Let $\varepsilon \in \mathbb{R}_o^+$. $x_o \in E$ is called ε-Kuhn-Tucker-point: $\Longleftrightarrow$

$$\bigvee_{\substack{\lambda \varepsilon F_1^* \\ \mu \varepsilon F_2^*}} \bigwedge_{y \varepsilon Y} \lambda(y-g(x_o)^+) \geq o \wedge \|f_x(x_o)-\lambda g_x(x_o)-\mu h_x(x_o)\| \leq \varepsilon$$

Every point fulfilling the generalized Euler-Lagrange-equalities in Dubovitskii-Milyutin-Theory is a o-Kuhn-Tucker-point [15, p.37] .

Theorem 16: $x_o \varepsilon E$ ε-Kuhn-Tucker-point and o-feasible for (P) $\Longrightarrow \bigwedge_{\alpha > \varepsilon} \bigvee_{r > o} x_o$ (α,r)-stationary

Proof: See Appendix.

Theorem 17: $x_o \varepsilon E$ ε-Kuhn-Tucker-point with Lagrange - multipliers λ,μ and σ_o-feasible, $\sigma_o > o$ and x_o regular with regularity number R and σ_1, α $\mathbb{R}_o^+$, $r \varepsilon \mathbb{R}^+$ satisfy the following inequalities

$$(\sigma_o R)^2 < r < \frac{\sigma_1}{C} \qquad (*)$$

$$(1 - \frac{\sigma_o R}{\sqrt{r}})\alpha \geq (\|\lambda\| + \|\mu\|)\frac{\sigma_o + \sigma_1}{\sqrt{r}} + \varepsilon + \|f_x(x_o)\| \frac{\sigma_o R}{\sqrt{r}} \qquad (**)$$

$\Longrightarrow$

$$\left(\bigvee_{n_o \in \mathbb{N}} \bigwedge_{n \geq n_o} x_n \ \sigma_1\text{-feasible}\right) \wedge x_o \ (\alpha,r)\text{-stationary}$$

Proof: See Appendix.

Theorem 18: $x_o \varepsilon E$, α, $r \varepsilon \mathbb{R}^+$, x_o (α,r)-stationary
$\Longrightarrow$ x_o α-Kuhn-Tucker-point.

Proof: See Appendix.

One can interpret Theorem 16,17,18 in the following way: If x_o is nearly feasible and regular then by theorem 16 and 17 x_o nearly Kuhn-Tucker-point implies x_o nearly stationary. By Theorem 18 x_o nearly stationary implies x_o nearly Kuhn-Tucker-point.

Now look at inequality (**) in particular to terms $\frac{\sigma_o}{\sqrt{r}}$ and $\frac{\sigma_1}{\sqrt{r}}$. If we can show that $\sigma_o/\sqrt{r}$ as well as $\sigma_1/\sqrt{r}$ can be made sufficiently small which means: in Theorem 17 α is close to ε, ε-Kuhn-Tucker-point and ε-stationary point is equivalent in the sense descripted in Theorem 16, 17 and 18. The assertion "$\sigma_o/\sqrt{r}$ and $\sigma_1/\sqrt{r}$ can be made as small as you want" is content of Lemma 19.

Lemma 19: $C \varepsilon \mathbb{R}_o^+$, $\sigma_o, R \varepsilon \mathbb{R}_o^+$ and $\sigma_o C R^2 < 1$

$$\Longrightarrow \bigvee_{(\sigma_i)_{i \in \mathbb{N}}} \bigvee_{(r_i)_{i \in \mathbb{N}}} \left(\bigwedge_{i \in \mathbb{N}} (\sigma_i R)^2 < r_i < \frac{\sigma_{i+1}}{C}\right) \wedge$$

$$\wedge \lim_{i \in \mathbb{N}} \frac{\sigma_i}{\sqrt{r_i}} = o$$

Proof: See Appendix

Now we can state as follows: If σ_i, r_i constructed as above than the solution of (LP_n^o) gives a criterion "as good" Kuhn-Tucker-condition is valid, and if other assumptions permit such conclusion a criterion for optimality. Another important question is: How strong is local descent $f_x(x_o)\ x_n$ if a stationary point is not even reached. The answer is given in the next theorem.

Theorem 20: $x_o \in E\ \sigma_o$-feasible and x_o regular with regularity number R and x_o is <u>not</u> ε-Kuhn-Tucker point and $r \in \mathbb{R}^+ \wedge$

$$R \frac{\sigma_o}{\sqrt{r}} < \frac{\varepsilon}{\varepsilon + \|f_x(x_o)\|} < 1 \qquad (***)$$

$$\Longrightarrow \bigwedge_{n \in \mathbb{N}} \Delta x_n \text{ solution of } (LP_n) \text{ exists } \wedge$$

$$\wedge\ f_x(x_o)\Delta x_n \leq R\sigma_o(\|f_x(x_o)\| + \varepsilon) - \sqrt{r}\varepsilon$$

$$\wedge\ f_x(x_o)\ \Delta x_n^\circ < -\ \varepsilon\sqrt{r}$$

Proof: See Appendix

Theorem 20 gives reason to a discussion analogous to that about theorem 16, 17, 18, as descent is guaranteed if $\sigma_o / \sqrt{r}$ is small enough. Using

$$f_x(x_o)\Delta x_n \leq (R \frac{\sigma_o}{\sqrt{r}} (\|f_x(x_o)\| + \varepsilon) - \varepsilon)\ \sqrt{r}$$

descent is guaranteed by (***).

Again application of Lemma 19 shows, that if $R, C, \|f_x(x_o)\|$ do not increase unlimited by practical use of Lemma 19 either a nearly stationary point must be reached of $f_x \Delta x_n$ is negative and by use of any line-search-procedure objective must decrease.

V. COMPUTATIONAL ASPECTS AND ALGORITHM

Although the basic concept is the same in chapter 3 and chapter 4, we treat the algorithm seperately. Only at the beginning of this chapter we shall make general remarks.

In general none of the real numbers C, R, L are known, and computation of these numbers is rather difficult. Only a guess of C, R, L can be used. There are two properties our parameter can have:
Either their values suffice that the algorithm work, or they do not.

On the other hand the algorithm can work or not.
The second case obviously shows that C, R, L is not sufficient for the working of the algorithm.

In practice the algorithms themselves are dependent on the guesses of C, R, L. We shall give underestimating approximation formulas for C, R, L:

$$C \geq \max\{g_2(s,x,\Delta x),\ h_2(s,x,\Delta x)\}$$

$$g_2(s,x,\Delta x) := \begin{cases} \dfrac{\|g(s,x+\Delta x)-g_x(s,x)\Delta x-g(s,x)\|}{\|\Delta x\|^2} & \text{if } \Delta x \neq 0 \\ 0 & \text{otherwise} \end{cases}$$

$h_2(s,x,\Delta x)$ analogous to $g_2(s,x,\Delta x)$

$$R \geq \begin{cases} \dfrac{\|\widetilde{\Delta x}_n\|}{\sigma(s,x)} & \text{if } \sigma(s,x) > 0 \\ 0 & \text{otherwise} \end{cases}$$

$$\sigma(s,x) := \max\{\|g(s,x)^-\|, \|h(s,x)\|\}$$

$\Delta\tilde{x}_n$ is norm-minimal feasible point for (LP_n)

$$L \geq \frac{\|\sigma(s+\Delta s,x)-\sigma(s,x)\|}{|\Delta s|} \quad \text{for} \quad \Delta s > 0$$

Now let us look for criterions to see which constant C, R, L has been underestimated too much:

(LP_n) has no solution for some n:
Obviously, by (*) and theorem 9 this indicates an underestimation of R. Our measure will be: increase R and try to choose σ_1,r again!
For all $n \in \mathbb{N}$ $x_o+\Delta x_n$ is not σ_1-feasible to (P). (In practical computation we only can try $n \in \{1,\ldots,n_o\}$.)
By theorem 7 and (*) this indicates underestimation of C. Now we increase C and choose σ_1,r again.

If the choice of σ_1,r is not possible with σ_1 limited by a fixed bound, we must try again from the beginning always using a smaller bound for σ_o, σ_1.

No problem occurs in estimating L as you can calculate $\sigma(s+\Delta x,x)$ and if $\sigma(s+\Delta s,x)$ overtops our bound we try a smaller Δs. Similar to the bound n_o for the series of problems (LP_n) we use such bounds for R, C too and if those bounds are exceeded the algorithms stop for reasons of complexity. Now we will give a flow-chart considering aspects mentioned above.

Algorithm solving "Find x $\bar{\sigma}$-feasible"

Given: $x_o, \bar{\sigma}, \sigma_{max}, C_{init}, R_{init}, \varepsilon, n_o$.

We look for a point x $\bar{\sigma}$-feasible to (P).

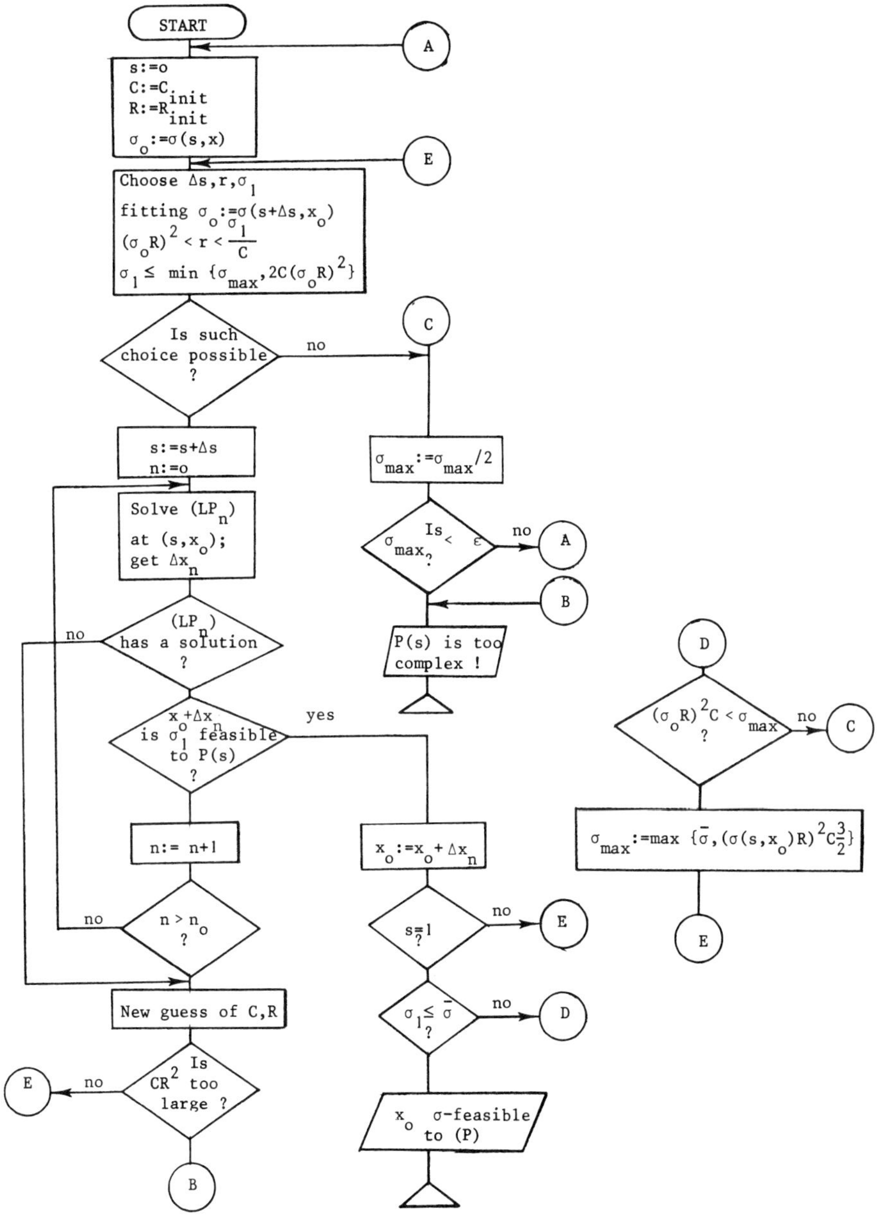

Algorithm solving "Find x α-stationary and x σ-feasible"

Given: $x_o, \bar{\sigma}, \bar{\alpha}, C_{init}, R_{init}, \varepsilon, n_o$.

We look for a point x $\bar{\sigma}$-feasible for (P) and $\bar{\alpha}$-stationary

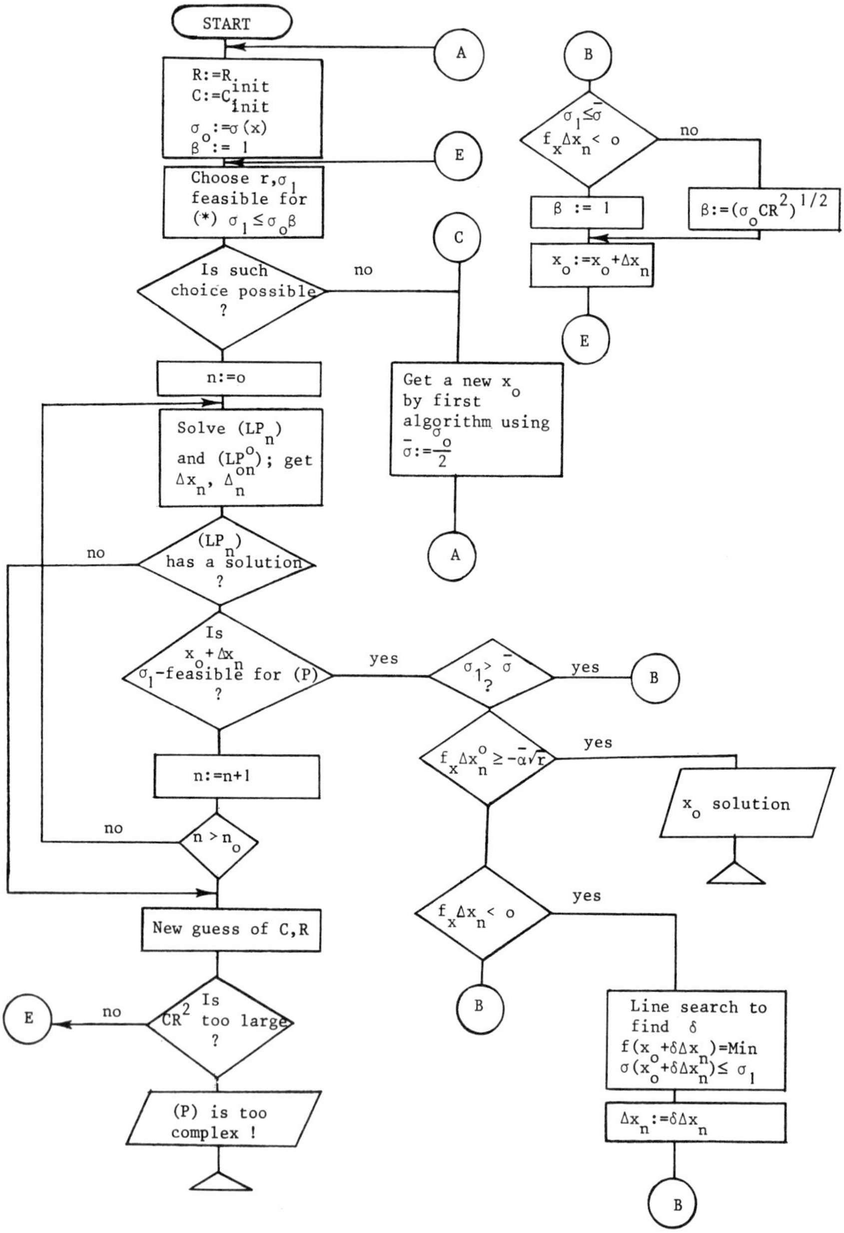

VI. SOLUTIONS OF LINEARIZED PROBLEMS (LP_N), (LP_N^0)

(LP_n) was defined in Definition 4:

$$\begin{cases} \langle f_x(x_o),\ \Delta x\rangle = \text{Min!} \\ \langle \Delta x,\ \Delta x\rangle \le r \\ \bigwedge\limits_{k\,\varepsilon\,\{1,\dots,n\}} \langle g(x_k)^-, g(x_o)+g_x(x_o)\Delta x - g(x_k)^+\rangle \le o \\ \bigwedge\limits_{k\,\varepsilon\,\{1,\dots,n\}} \langle h(x_k), h(x_o)+h_x(x_o)\Delta x\rangle = o \end{cases}$$

Using the following abbreviations:

$$\begin{aligned} f_k &:= h_x(x_o)^{ad} \circ h(x_k) && k=1,\dots,n \\ f_{n+k} &:= g_x(x_o)^{ad} \circ g(x_k)^- && k=1,\dots,n \\ f_{2n+1} &:= f_x(x_o) && \\ c_k &:= -\langle h(x_k), h(x_o)\rangle && k=1,\dots,n \\ c_{k+n} &:= -\langle g(x_k)^-, g(x_o)-g(x_k)^+\rangle && k=1,\dots,n \end{aligned}$$

we can write (LP_n):

$$\begin{cases} \langle \Delta x, \Delta x\rangle \le r \\ \bigwedge\limits_{k\,\varepsilon\,\{1,\dots,n\}} \begin{array}{l} \langle f_k, \Delta x\rangle = c_k \\ \langle f_{k+n}, \Delta x\rangle \le c_{k+n} \end{array} \\ \langle f_{2n+1}, \Delta x\rangle = \text{Min!} \end{cases}$$

An easy proof shows, that it is sufficient to look for solutions Δx only in the linear hull of $\{f_1,\dots,f_{2n+1}\}$. By

$$\Delta x := \sum_{i=1}^{2n+1} \lambda_i f_i$$

and $a_{ij} := \langle f_i, f_j \rangle$ (LP_n) gets the following shape

$$\begin{cases} \sum_{j=1}^{2n+1} a_{ij}\lambda_j \begin{cases} = c_i & \text{for } i \in \{1,\dots,n\} \\ \le c_i & \text{for } i \in \{n+1,\dots,2n\} \\ = \text{Min!} & \text{for } i = 2n+1 \end{cases} \\ \sum_{i=1}^{2n+1} \sum_{j=1}^{2n+1} \lambda_i a_{ij} \lambda_j \le r \end{cases}$$

or shorter

$$\begin{cases} A\ \lambda \begin{cases} = & c \\ \le & c \\ = & \text{Min!} \end{cases} \\ \lambda^T A\ \lambda \le r \end{cases} \qquad (LP_n')$$

We don't want to treat (LP_n') here because there is no importance how we get the solution, because we are only interested in the solution itself. How to solve (LP_n') is treated in [21] respectively [22].

It is trivial to see, that (LP_n^o) also leads to (LP_n') only with c=o. If objective $\langle f_x(x_o), \Delta x \rangle$ is replaced by $\langle \Delta x, \Delta x \rangle$ we get

$$A\lambda \begin{cases} = & c \\ \le & c \end{cases}$$

$$\lambda^T A \lambda = \text{Min!}$$

which is a problem often treated in literature.

APPENDIX A (PROOFS)

Proof of Lemma 5:

As Y is convex, the feasible region of (LP) is included in the feasible region of (LP_{k-1}), which proves the first proposition. Now if (LP) has a feasible point (LP_{k-1}) also has one. We only have to show that in this case (LP_{k-1}) has an optimal solution. Chapter 6 shows that (LP_{k-1}) can be reduced to a simply solvable finite dimensional problem.

Proof of Lemma 6:

By definition of a^+, b^+, a^- and [23, p.69]

$$\langle b^-, b^+ - a^+ \rangle \geq 0 .$$

Now

$$\langle a^-, a^+ - b^+ \rangle \leq \langle a^- - b^-, a^+ - b^+ \rangle = \frac{1}{4} \{ \| (a^- - b^-) + (a^+ - b^+) \|^2 - \| (a^- - b^-) - (a^+ - b^+) \|^2 \} \leq \frac{1}{4} \| a - b \|^2$$

and

$$\| a^- - b^- \|^2 + \| a^+ - b^+ \|^2 \leq \| a^- - b^- \|^2 + 2\langle a^-, a^+ - b^+ \rangle - 2\langle b^-, a^+ - b^+ \rangle + \| a^+ - b^+ \|^2 = \| (a^- - b^-) + (a^+ - b^+) \|^2 = \| a - b \|^2 .$$

Proof of theorem 7:

I) g,h compact (i)

$K := \{ \Delta x \in E \,|\, \langle \Delta x, \Delta x \rangle \leq r \}$ bounded (ii)

Let $(y_n)_{n \in \mathbb{N}}$ be any sequence in K and t one of the mappings g,h.

(i) and (ii) implies

$$\bigwedge_{\delta > 0} \; \bigvee_{n_o \in \mathbb{N}} \; \bigwedge_{n \geq n_o} \; \bigvee_{k < n} \|t(y_k) - t(y_n)\| \leq \delta \qquad \text{(iii)}$$

II) Let $\varepsilon \in \mathbb{R}^+$

$\delta := \varepsilon/2$ and $k,n \in \mathbb{N}$ be such that $k < n$ and

$$\|g(x_k) - g(x_n)\| \leq \delta \qquad \text{(iv)}$$

Now the following inequalities and equalities are valid:

$\|g(x_n)^-\|^2 = (A) + (B) + (C) + (D) + (E)$

with

$(A) := \langle g(x_k)^-, g(x_o) + g_x(x_o)\Delta x_n - g(x_k)^+ \rangle \leq o$

by definition 4 and $k<n$.

$(B) := \langle g(x_k)^-, g(x_k)^+ - g(x_n)^+ \rangle \leq (\frac{\delta}{2})^2$

by Lemma 6 and (iv).

$(C) := \langle g(x_n)^-, g(x_n) - g(x_o) - g_x(x_o)\Delta x_n \rangle \leq \|g(x_n)^-\| \, Cr$

by [DIF].

$(D) := \langle g(x_n)^- - g(x_k)^-, g(x_n)^- \rangle \leq \delta \, \|g(x_n)^-\|$

by (iv).

$(E) := -\langle g(x_n)^- - g(x_k)^-, g(x_n) - g(x_o) - g_x(x_o)\Delta x_n \rangle \leq \delta \, Cr$

by (iv) and [DIF].

Altogether

$\|g(x_n)^-\|^2 - \|g(x_n)^-\|(\delta+Cr) - (\frac{\delta}{2})^2 - \delta Cr \leq o$

and therefore

$\|g(x_n)^-\| \leq \quad Cr+2\delta = Cr+\varepsilon$ (v)

Analogous calculations lead to

$\|h(x_n)\| \quad \leq \quad Cr+\varepsilon$ for sufficiently large n.

and by (iv) and an analogous statement for h we get

$\bigvee_{n_o \in \mathbb{N}} \quad \bigwedge_{n \geq n_o} \quad x_o+\Delta x_n$ $(\varepsilon+Cr)$-feasible for (P).

III) Now we show

Δx_n is $(Cr+\varepsilon)$-feasible for (LP) if

$\|g(x_n)-g(x_k)\| \leq \delta \wedge \|h(x_n)-h(x_l)\| \leq \delta$ for some $k,l<n$

Let $\|g(x_n)-g(x_k)\| \leq \delta$

By [DIF]:

$\|g(x_o)+g_x(x_o)\Delta x_n-g(x_k)\| \quad \leq Cr+\delta$

Using $g(x_k) = g(x_k)^+ + g(x_k)^-$ and Definition 4 we get from

$\|g(x_o)+g_x(x_o)\Delta x_n-g(x_k)^+-g(x_k)^-\|^2 \leq (Cr+\delta)^2$:

$\|g(x_o)+g_x(x_o)\Delta x_n-g(x_k)^+\|^2 + \|g(x_k)^-\|^2$

$-2\langle g(x_o)+g_x(x_o)\Delta x_n-g(x_k)^+, g(x_k)^-\rangle \leq (Cr+\delta)^2,$

$\|(g(x_o)+g_x(x_o)\Delta x_n)^-\|^2 + \|(g(x_o)+g_x(x_o)\Delta x_n)^+-g(x_k)^+\|^2$

$+2\langle(g(x_o)+g_x(x_o)\Delta x_n)^-, (g(x_o)+g_x(x_o)\Delta x_n)^+-g(x_k)^+\rangle \leq (Cr+\delta)^2,$

By [23,p.69] and definition of a^+, a^-

$\|(g(x_o)+g_x(x_o)\Delta x_n)^-\| \leq \quad Cr+\delta$

Analogous calculations for $h(x_o)+h_x(x_o)\Delta x_n$ prove the second part of the theorem.

Proof of Theorem 12:

By definition of $\underline{a}$, $\overline{a}$ for all real numbers $q \in (\underline{a},\overline{a})$ and pairs $(s,x) \in \Gamma$

$$r(s,x) < q/C(s,x)$$

and $$q^2R(s,x)^2 < r(s,x)$$

Now we define $\sigma := \min\{ q,\overline{\sigma} \}$ after choosing $q \in (\underline{a},\overline{a})$ and $\tilde{r}(s,x) := r(s,x)\,.\,\frac{\sigma}{q}$ for all $(s,x) \in \Gamma$

Obviously the following inequalities hold for $(s,x) \in \Gamma$

$$\tilde{r}(s,x) < \sigma/C(s,x)$$

and $$\sigma^2R(s,x)^2 < \tilde{r}(s,x)$$

Now let $d \in (o,1)$ and for all pairs $(s,x) \in \Gamma$ choose $\sigma_o := \sigma d^2$, $\sigma_1 := \sigma d$, $r := \tilde{r}d$, $\Delta s := \sigma d(1-d)/L$.

$m :=$ smallest integer greater or equal to ${}^1/\Delta s$

$s_i := i\Delta s \quad i=0,1,\ldots,m-1$ and $s_m := 1$

Now by induction we show the following proposition:

$$\bigwedge_{i \in \{1,\ldots,m\}} \ \bigvee_{k_i \in \mathbb{N}} \begin{array}{l} x^i-x^{i-1} \text{ is solution of } (LP_{k_i}) \text{ at } (s_i,x^{i-1}) \\ \text{and } x^i \text{ is } \sigma\frac{a}{q}\text{-feasible for } (P(s_i)) \\ \text{if we choose } r := \tilde{r}(s_i,x^{i-1}) \end{array}$$

Always $r := \tilde{r}(s_i,x^{i-1}) \leq r(s_i,x^{i-1})$.

First x^o is 0-feasible for $(P(o))$ and $L\Delta s$-feasible for $(P(s_1))$

$L\Delta s = \sigma\beta - \sigma\beta^2 < \sigma$.

Now by $\sigma_o^2R(s_i,x^{i-1})^2 = \sigma^2d^4R(s_i,x^{i-1})^2 < \tilde{r}(s_i,x^{i-1})d = r < \sigma d/c = = \sigma_1/c$ (*) in Theorem 1o holds guaranteeing (s_i,x^i) being σ_1-feasible if (s_i,x^{i-1}) is σ_o- feasible for $(P(s_i))$ x^o is o-feasible for $(P(o))$ and by induction and $\sigma_1 + L\Delta s \leq \sigma_o$ (s_i,x^i) is σ_o-feasible for all $i \in \{o,\ldots,m\}$.

Especially by $\sigma_1 \leq \bar{\sigma}$ x^m is $\bar{\sigma}$-feasible for (P(1)).

Proof of Theorem 16:

x_o ε-Kuhn-Tucker-point means

$$\bigvee_{\lambda \in F_1^*,\ \mu \in F_2^*} \lambda \in (Y-g(x_o)^+)^* := \{\lambda \in F_1^* / \bigwedge_{y \in Y} \lambda(y-g(x_o)^+) \geq o\}$$

$$\wedge \|f_x(x_o) - \lambda g_x(x_o) - \mu h_x(x_o)\| \leq \varepsilon$$

Choose $r \in \mathbb{R}^+$ satisfying

$$\sqrt{r} < \alpha/(C(\|\lambda\| + \|\mu\|)) \qquad \text{(i)}$$

If λ, μ equal zero the proposition is trivial. As x_o is 0-feasible x_o is regular and by theorems 9, 10 (LP_n) has a solution. (LP_n^o) is well defined now.

As $\langle g(x_n)^-, g(x_o)-g(x_n)^+\rangle \leq o$

each point feasible for (LP_n^o) is feasible for (LP_n).

Let Δx_n^o be the optimal solution of (LP_n^o) and Δx_n the optimal solution of (LP_n). Then the following equalities and inequalities are valid:

$$f_x(x_o)\Delta x_n^o \geq f_x(x_o)\Delta x_n \geq \lambda g_x(x_o)\Delta x_n + \mu h_x(x_o)\Delta x_n - \varepsilon\sqrt{r} \geq$$

$$\geq \lambda((g(x_o)+g_x(x_o)\Delta x_n)^+ - g(x_o)) + \lambda(g(x_o)+g_x(x_o)\Delta x_n)^- +$$

$$+ \mu(h(x_o) + h_x(x_o)\Delta x_n) - \varepsilon\sqrt{r} \geq$$

$$\geq -\|\lambda\|\sigma_1 - \|\mu\|\sigma_1 - \varepsilon\sqrt{r}$$

For all $\delta \in \mathbb{R}^+$ equality $Cr + \delta =: \sigma_1$ and theorem 7 and $n \in \mathbb{N}$ large enough the last inequality holds.

Now

$$f_x(x_o)\Delta x_n^o \geq -(\|\lambda\|+\|\mu\|)(Cr+\delta) - \varepsilon\sqrt{r}$$

δ sufficiently small and (i) shows

$$f_x(x_o)\Delta x_n^o \geq -(\alpha+\varepsilon)\sqrt{r}$$

Proof of theorem 17:

By theorem 9 (LP_n) and (LP_n^o) is well defined for all integers n. Let n_o be one of those integers, such that for all larger integers n x_n is σ_1-feasible for (P) and $\Delta x_n := x_n - x_o$ is σ_1-feasible for (LP). We want to show

x_o is (α,r)-stationary

which means $f_x(x_o)\Delta x_n^o \geq -\alpha\sqrt{r}$ (i)

when Δx_n^o is solution of (LP_n^o).

Using the assumptions of the theorem the following inequalities are valid.

$$f_x(x_o)\Delta x_n \geq \lambda g_x(x_o)\Delta x_n + \mu h_x(x_o)\Delta x_n - \varepsilon\|\Delta x_n\| \geq$$

$$\geq \lambda((g(x_o) + g_x(x_o)\Delta x_n)^+ - g(x_o)^+) + \lambda((g(x_o) + g_x(x_o)\Delta x_n)^-)$$

$$- \lambda g(x_o)^- + \mu (h(x_o) + h_x(x_o) \Delta x_n) - \mu h(x_o) - \varepsilon\sqrt{r} \geq$$

$$\geq 0 - \|\lambda\| \sigma_1 - \|\lambda\| \sigma_o - \|\mu\| \sigma_1 - \|\mu\| \sigma_o - \varepsilon\sqrt{r} \geq$$

$$\geq - (\|\lambda\| + \|\mu\|)(\sigma_o + \sigma_1) - \varepsilon\sqrt{r} \qquad \text{(ii)}$$

Next we want to find a lower bound to $f_x(x_o)(\Delta x_n^o - \Delta x_n)$. Let $\Delta\tilde{x}$ be the normminimal feasible point for (LP_n). Obviously $\|\Delta\tilde{x}\| \leq R\sigma_o$. Now

$$\begin{aligned} f_x(x_o)\Delta x_n &= \inf\{ f_x(x_o)\Delta x \mid \Delta x \text{ feasible for } (LP_n)\} \\ &\leq \inf\{ f_x(x_o)(\Delta x+\Delta\tilde{x}) \mid \Delta x \text{ feasible for } (LP_n^o) \text{ replacing } \\ &\qquad \|\Delta x\|^2 \leq r \text{ by } \|\Delta x+\Delta\tilde{x}\|^2 \leq r \} \\ &\leq f_x(x_o)\Delta\tilde{x}+\inf\{f_x(x_o)\Delta x \mid \Delta x \text{ feasible for } (LP_n^o) \\ &\qquad \text{replacing } \|\Delta x\|^2 \leq r \text{ by } \\ &\qquad \|\Delta x\|^2 \leq (\sqrt{r} - \|\Delta\tilde{x}\|)^2 \} \\ &= f_x(x_o)\Delta\tilde{x} + \frac{\sqrt{r}- \|\Delta\tilde{x}\|}{\sqrt{r}} f_x(x_o)\Delta x_n^o \end{aligned} \qquad \text{(iii)}$$

By (iii)

$$f_x(x_o)(\Delta x_n^o - \Delta x_n) \geq -f_x(x_o)\Delta\tilde{x}+f_x(x_o)\Delta x_n^o \frac{\|\Delta\tilde{x}\|}{\sqrt{r}}$$

(ii) (iii) and the above inequality imply

$$f_x(x_o)\Delta x_n^o = f_x(x_o)\Delta x_n + f_x(x_o)(\Delta x_n^o - \Delta x_n) \geq -(\|\lambda\|+\|\mu\|)(\sigma_o+\sigma_1) -$$

$$- \varepsilon\sqrt{r} - f_x(x_o)\Delta\tilde{x} + \frac{\|\Delta\tilde{x}\|}{\sqrt{r}} f_x(x_o)\Delta x_n^o$$

Further

$$f_x(x_o)\Delta x_n^o \geq -(1-\|\Delta\tilde{x}\|/\sqrt{r})^{-1}((\|\lambda\| + \|\mu\|)(\sigma_o+\sigma_1)/\sqrt{r} +$$

$$+ \varepsilon + f_x(x_o)\Delta\tilde{x}/\sqrt{r})\sqrt{r} \geq -(1-\sigma_o R/\sqrt{r})^{-1}((\|\lambda\| + \|\mu\|)(\sigma_o+\sigma_1)/\sqrt{r} +$$

$+ \varepsilon + \|f_x(x_o)\| \, \sigma_o R/\sqrt{r})\sqrt{r} \geq - \alpha\sqrt{r}$

Proof of theorem 18:

Let x_o be (α,r)-stationary and let

$$\delta_n := \inf \{f_x(x_o)\Delta x/\sqrt{r} \mid \Delta x \text{ feasible for } (LP_n^o)\}$$

Obviously

$$\bigvee_{n_o \in \mathbb{N}} \ \bigwedge_{n \geq n_o} \delta_n \geq -\alpha \tag{i}$$

Let

$$S_n^* := \{ \sum_{i=1}^{n} \lambda_i g_x(x_o)^{ad} g(x_i)^- + \mu_i h_x(x_o)^{ad} h(x_i) \mid \lambda_i \in \mathbb{R}_o^-, \ \mu_i \in \mathbb{R}\}$$

$$S_n := \{ \Delta x \in E \mid \bigwedge_{v \in S_n^*} v\Delta x \geq 0 \}$$

(The definition of dual cones and elementary relations between dual cones see [15])

$$T^* := \{ g_x(x_o)^{ad} \circ \lambda + h_x(x_o)^{ad} \circ \mu \mid \lambda \in (Y-g(x_o)^+)^* \ , \quad \mu \in F_2^* \}$$

$$T := \{ \Delta x \in E \mid \bigwedge_{v \in T^*} v\,\Delta x \geq 0 \}$$

Obviously $T^* \supset S_n^*$ and $S_n \supset T$ for all $n \in \mathbb{N}$.

Now by the theorem of Mazur or a corollary [23, p.136]

$$\inf \{ \| f_x(x_o) - \lambda g_x(x_o) - \mu h_x(x_o) \| \;/\; \lambda \in (Y - g(x_o))^*, \quad \mu \in F_2^* \}$$

$$= d(f(x_o), T^*) \le d(f_x(x_o), S_n^*) = -\delta_n \le \alpha$$

Proof of lemma 19:

Let $\beta := (\sigma_o C R^2)^{\frac{1}{2}} < 1$ and $\bigwedge_{n \in \mathbb{N}} \sigma_n := \beta^n \sigma_o$

Now $(\sigma_n R)^2 = \beta^{2n} \sigma_o^2 R^2 = \dfrac{\sigma_o (\beta^{n+1})^2}{C} = \dfrac{\sigma_{n+1} \beta^{n+1}}{C} < \dfrac{\sigma_{n+1}}{C}$

Further let

$$\bigwedge_{n \in \mathbb{N}} r_n := \beta^{\frac{1}{2}} \frac{\sigma_{n+1}}{C}$$

Inequality $r_n < \dfrac{\sigma_{n+1}}{C}$ holds. Further

$$\bigwedge_{n \in \mathbb{N}} (\sigma_n R)^2 / r_n = \sigma_{n+1} \beta^{n+1} C / (C \sigma_{n+1} \beta^{\frac{1}{2}}) = \beta^{n+\frac{1}{2}} < 1$$

Now proposition

$$(\sigma_n R)^2 < r_n < \frac{\sigma_{n+1}}{C}$$

is proved.

$$0 \le \frac{\sigma_n}{\sqrt{r_n}} = (\sigma_o C)^{\frac{1}{2}} \beta^{\frac{n}{2} - \frac{3}{4}} \quad \text{tends to zero.}$$

Now

$$\lim_{n \in \mathbb{N}} \frac{\sigma_n}{\sqrt{r_n}} = 0 \text{ is proved.}$$

Proof of theorem 20:

Some arguments as in proof of theorem 17 show that (LP_n) has a solution and for each $n \in \mathbb{N}$ there exists a point $\Delta\tilde{x}, \|\Delta\tilde{x}\| \leqq R\sigma_o$ and $\Delta\tilde{x}$ feasible for (LP_n).

x_o is not ε-Kuhn-Tucker-point and definition 15.

$$\bigvee_{\substack{\Delta x \in E \\ \|\Delta x\| = 1}} \quad \bigwedge_{\substack{\lambda \in (y-g(x_o)^+)^* \\ \mu \in F_2}} \quad \begin{array}{l} \lambda g_x(x_o)\Delta\bar{x} \geq 0 \\ \mu h_x(x_o)\Delta\bar{x} = 0 \\ f_x(x_o)\Delta\bar{x} < -\varepsilon \end{array}$$

Now for fixed $\Delta\bar{x}$, $\Delta\tilde{x}$ the following proposition holds

$$\bigwedge_{\delta \geq 0} \delta\,\Delta\bar{x} \text{ is 0-feasible for } (LP_n^o) \text{ omitting } \langle\Delta x,\Delta x\rangle \leq r$$

$$\bigwedge_{\delta \geq 0} \Delta\tilde{x} + \delta\Delta\bar{x} \text{ is 0-feasible for } (LP_n) \text{ omitting } \langle\Delta x,\Delta x\rangle \leq r$$

Now we fix $\delta := \sqrt{r} - \|\Delta\tilde{x}\| > 0$ and get

$$\begin{aligned} f_x(x_o)\Delta x_n &\leq f_x(x_o)(\Delta\tilde{x}+\delta\Delta\bar{x}) \\ &\leq \|f_x(x_o)\| \; \|\Delta\tilde{x}\| - \delta\varepsilon \\ &\leq R\sigma_o\|f_x(x_o)\| - (\sqrt{r} - R\sigma_o)\varepsilon \\ &\leq R\sigma_o(\quad \|f_x(x_o)\| + \varepsilon) - \sqrt{r}\,\varepsilon \end{aligned}$$

Inequality

$$f_x(x_o)\Delta x_n^o < -\varepsilon\sqrt{r}$$

follows immediately from definition 14 and theorem 18.

REFERENCES

[1] Anonymous, A new algorithm for Optimization, Math. Programming 3 (1972), 124-128

[2] Blankenship, J. W., Falk, J.E., Infinitely Constrained Optimization Problems, JOTA 19 (1976), 261-281

[3] Fiacco, A. V., McGormick, G. P., Sequential unconstrained minimization techniques for nonlinear programming, a primal-dual method, Management Sci. 10, Nr. 2 (1964)

[4] Fiacco, A. V., A general regularized sequential unconstrained minimization technique, SIAM J. Appl. Math. 17 (1969), 1239-1245

[5] Gill, P. E., Murray, W., Numerical Methods for Constrained Optimization, Academic Press, New York, London (1974)

[6] Girsanov, I. V., Lectures on Mathematical Theory of Extremum Problems, Springer-Verlag, Berlin-New York (1972)

[7] Gould, F. J., Extensions of Legrange multiplies in nonlinear programming, SIAM J. Appl. Math. 17 (1969), 1280-1297

[8] Hackl, J., Zerlegung positiv definiter Matrizen und Anwendungen in der Optimierung, erscheint in ZAMM, Tagungsbericht 1977

[9] Hackl, J., Tobisch, H., Ein endliches Gradientenverfahren zur Lösung eines linearisierten Optimierungsproblems mit euklidischer Normbeschränkung, Inst. Ber. Nr. 89, Univ. Linz

[10] Hestenes, M. R., Optimization Theory, The finite dimensional case, John Wiley & sons, New York-London (1975)

[11] Lempio, F., Tangentialmannigfaltigkeiten und infinite Optimierung, Habilitationsschrift, Hamburg (1972)

[12] Lootsma, F. A., Numerical methods for nonlinear Optimization, Academic Press, London-New York (1972)

[13] Luenberger, D. G., Optimization by vector space methods, John Wiley & Sons, Inc., New York-London (1968)

[14] Pietrzykowski, T., An exact potential method for constrained maxima, J. SIAM Num. Anal. 6 (1969), 299-304

[15] Robinson, S. M., A quadratically convergent algorithm for general nonlinear programming problems, Math. Programming 3 (1972), 145-156

[16] Robinson, S. M., Extension of Newton's method for nonlinear functions with values in a cone, Num. Math. 19 (1972), 341-347

[17] Rosen, J. B., The Gradient Projection Method for Nonlinear Programming, Part III, Nonlinear Constraints, J. SIAM 9 (1961), 514-532

[18] Schittkowski, K., Numerical Solution of Systems of Nonlinear Inequalities in Optimization and Operations Research, Prof. Oberwolfach 1975, Springer-Verlag, New York 1976

[19] Tapia, R. A., The Differentiation and Integration of Nonlinear Operators in Nonlinear functional analysis and applications, Academic Press, London-New York, 1971

[20] Wacker, H. J., Ein Iterationsverfahren zur Lösung spezieller nichtlinearer Randwertprobleme, Computing 8 (1972), 275-291

[21] Yoshida, K., Functional Analysis, Springer-Verlag, Berlin-New York, 1974

[22] Zangwill, W. I., Nonlinear programming via penalty function, Management Sci. 13 (1967), 344-358

[23] Zoutendijk, Methods of Feasible Directions, Elsevier Pub. Co., 1960

SOME REMARKS ON THE CONVERGENCE OF DESCENT METHODS

G. Heindl

Technische Universität München
Institut für Mathematik
Munich, Federal Republic of Germany

0. INTRODUCTION

The problem of solving a nonlinear operator equation $F(x)=0$ can often be treated very effectively by converting it into the problem of minimizing a suitable functional f. If for instance F is a C^1-mapping from an open subset G of $\mathbb{R}^n$ to $\mathbb{R}^n$, a common choice for f is

$$f : G \ni x \to \frac{1}{2} F(x)^T F(x) \in \mathbb{R} .$$

Let us use this special case for our introductory considerations.

ISBN 0-12-729250-0

Clearly the equation $F(x)=0$ has a solution if and only if there is a minimizer for f and the minimal value of f is zero. In order to assure the existence of a solution of $F(x)=0$, one generally assumes there is a compact connected or path-connected component L^o of a level set

$$L(f(x^o)) := \{x \in G \mid f(x) \leq f(x^o)\}$$

of f such that $F'(x)$ is regular whenever $x \in L^o$. It is well known that under this assumption the subset of zeros of F contained in L^o is nonempty and finite. As a consequence, for every descent method producing a sequence $\{x^k\}$ in L^o with the properties

a) $\lim\limits_{k\to\infty} f(x^k) = 0$ and

b) $\lim\limits_{k\to\infty} (x^{k+1} - x^k) = 0,$

this sequence converges to a zero of F in L^o. In fact it can be shown (Theorem 1) that there is exactly one zero of F in L^o and therefore every sequence in L^o with the properties a) and b) must have the same limit. This result is derived from some more general properties of level sets, studied in the first section. As an application, we give a convergence theorem for a damped Newton method and a continuation method derived from it (Theorems 2 and 3).

1. SOME PROPERTIES OF LEVEL SETS

In what follows let f denote a real valued C^1-function defined on an open subset G of $\mathbb{R}^n$. We are mainly interested in the behaviour of f on compact connected or path-connected components of its level sets $L(f(x^o)) := \{x \in G \mid f(x) \le f(x^o)\}$, $x^o \in G$. First it will be shown that under natural assumptions it is not necessary to distinguish between compact connected and compact path-connected components of level sets.

LEMMA 1. Let $x^o \in G$ and $L^o \subset L(f(x^o)) =: L$.
Assume that L^o is closed with respect to L (or G) and that $f'(x) \neq 0$ whenever $x \in L^o$ and $f(x) = f(x^o)$. Then L^o is a connected component of L iff it is a path-connected component of L.

We shall prove this Lemma as a Corollary of

PROPOSITION 1. If $f'(\bar{x}) \neq 0$, then $\bar{x}$ has a path-connected neighbourhood with respect to $L(f(\bar{x}))$.

Proof of Prop. 1. Let $f'(\bar{x}) \neq 0$.
Set $E := \{h \in \mathbb{R}^n \mid f'(\bar{x})(h) = 0\}$ and fix a $p \in \mathbb{R}^n$ with $f'(\bar{x})(p) = 1$. Then, given any norm $\|.\|$ on $\mathbb{R}^n$, there exist $\alpha, \beta > 0$ such that $\bar{x} + h + tp \in G$ whenever

$$(h,t) \in U := \{(h',t') \in E \times \mathbb{R} \mid \|h'\| < \alpha, |t'| < \beta\}.$$

Now consider the C^1-function $\varphi: U \to \mathbb{R}$ defined by

$$\varphi(h,t) := f(\bar{x} + h + tp) - f(\bar{x}).$$

From $\varphi(0,0)=0$ and $D_2\varphi(0,0) = id_{\mathbb{R}}$ we conclude, applying the implicit function theorem:

There exist $\alpha', \beta' > 0$ and a real valued C^1-function g defined on

$$W_1 := \{h \in E \mid \|h\| < \alpha'\}$$

such that $\alpha' \leq \alpha$, $\beta' \leq \beta$ and for any $h \in W_1$, $t=g(h)$ is the unique solution of the equation $\varphi(h,t)=0$ in

$$W_2 := \{t' \in \mathbb{R} \mid |t'| < \beta'\}.$$

$V := \{\bar{x} + h + tp \mid h \in W_1, \; t \in W_2\}$ is a neighbourhood of $\bar{x}$ (with respect to $\mathbb{R}^n$) and any $x = \bar{x} + h + tp \in V \cap L(f(\bar{x}))$ can be joined in V $L(f(\bar{x}))$ with $\bar{x}$ using the path γ defined by

$$\gamma(s) := \begin{cases} \bar{x} + 2sh + g(2sh)p \text{ when } o \leq s \leq \frac{1}{2} \\ \bar{x} + h + [(2-2s)g(h) + (2s-1)t]p \text{ when } \frac{1}{2} \leq s \leq 1. \end{cases}$$

Thus $V \cap L(f(\bar{x}))$ is a path-connected neighbourhood of $\bar{x}$ with respect to $L(f(\bar{x}))$.

Remark. Clearly $f'(\bar{x}) \neq 0$ also implies that $\bar{x}$ has a path-connected neighbourhood with respect to the set $\{x \in G \mid f(x) \geq f(\bar{x})\}$.

Proof of Lemma 1.

Every $x \in L^o$ has a path-connected neighbourhood $U(x)$ with respect to L. This is clear for all $x \in L^o$ with $f(x) < f(x^o)$ and follows from Prop. 1 for all $x \in L^o$ with $f(x)=f(x^o)$. If L^o is a connected component of L, we have $U(x) \subset L^o$ for every $x \in L^o$. This implies that L^o is path-connected and in fact a path-connected component of L (cf. [4] p. 116 Th. 5.5). If L^o is a path-connected component of L then we also have $U(x) \subset L^o$ for every $x \in L^o$. Thus L^o is open with respect to L. Let $\tilde{L}$ denote the connected component of L containing L^o. L^o being both open and closed with respect to L, we deduce that $L^o = \tilde{L}$.

Most of our later results follow from

LEMMA 2. Let L^o be a compact connected or path-connected component of $L := L(f(x^o))$ and let $x^* \in L^o$ be such that $f(x^*) < f(x^o)$ and $f'(y) \neq 0$ whenever $y \in L^o$ and $f(y) > f(x^*)$. Then $Q := L^o \cap L(f(x^*))$ is a connected component of $L(f(x^*))$.

The proof of this assertion is based on a construction similar to one used in [1] to show that every C^1-function $F: \mathbb{R}^n \to \mathbb{R}$ satisfying the growth condition $F(x) \to \infty$ as $\|x\| \to \infty$ and having two isolated relative minima, must have a further critical point ([1] p. 57 2-25).

For later purposes we present the main steps of the proof in three propositions.

PROPOSITION 2. Let $\|\,.\,\|$ denote the euclidean norm on $\mathbb{R}^n$ and y a point in G such that $f'(y) \neq 0$. Then for every sufficiently small positive $r \in \mathbb{R}$ there exists a continuous mapping $Ty^r : G \to G$ with the properties

(I) $x \in G$, $\| Ty^r(x) - y \| < r$ and $f(Ty^r(x)) < f(x)$ whenever $\|x-y\| < r$,

(II) $Ty^r(x) = x$ for all $x \in G$ with $\|x-y\| \geq r$.

The proposition can be proved by a well known perturbation technique. We give the short proof for the sake of completeness.

For every sufficiently small positive $r \in \mathbb{R}$ we have $x \in G$ and $\| \operatorname{grad} f(x) \| < 2 \| \operatorname{grad} f(y) \| =: \beta$ whenever $\|x-y\| < r$, while

$$\operatorname{grad} f(x_1)^T \operatorname{grad} f(x_2) > 0$$

whenever $\| x_i - y \| < r$, $i=1,2$.

Now Ty^r can be defined by

$$Ty^r(x) := \begin{cases} x - \beta^{-1}(r - \|x-y\|) \operatorname{grad} f(x) & \text{when } \|x-y\| < r \\ x & \text{for all other } x \in G \end{cases}$$

In fact, $\| x-y \| < r$ implies

$$\|Ty^r(x) - y\| \leq \| Ty^r(x) - x \| + \|x-y\| = \beta^{-1} \|\operatorname{grad} f(x)\| (r - \|x-y\|) + \| x-y \| < r$$

from which we derive the existence of a $\bar{x} \in G$ such that $\|\bar{x}-y\| < r$ and $f(Ty^r(x))-f(x) = (\text{grad } f(\bar{x}))^T(Ty^r(x)-x) = -\beta^{-1}(r-\|x-y\|)(\text{grad } f(\bar{x}))^T\text{grad } f(x) < 0$.

PROPOSITION 3. Let Q, K be disjoint subsets of G, Q closed with respect to G, K compact, nonempty and such that $f'(y) \neq 0$ for all $y \in K$.

Then there is an open set U and a continuous mapping $T: G \to G$ with the properties

(I) $K \subset U \subset G \setminus Q$,

(II) $T(x) \in U$ and $f(T(x)) < f(x)$ for all $x \in U$,

(III) $T(x) = x$ for all $x \in G \setminus U$.

Proof. For every $y \in K$ there is a positive $r(y) \in \mathbb{R}$ and a continuous mapping $Ty^{r(y)}: G \to G$ such that $\|x-y\| \geq r(y)$ for all $x \in Q$ and (I) and (II) of Prop. 2 hold with r replaced by r(y). Since K is compact, there exist $y^1, \ldots, y^k \in K$ such that $U := \{x \in \mathbb{R}^n \mid \|x-y^i\| < r(y^i) \text{ for some } i \in \{1,\ldots,k\}\}$ contains K. Now it can easily be shown by induction that T can be taken to be the product

$$Ty_k^{r(y_k)} \circ \ldots \circ Ty_1^{r(y_1)}.$$

PROPOSITION 4. Let L^o be a compact connected component of $L := L(f(x^o))$ and Q a nonempty compact subset of L^o such that $f'(y) \neq 0$ for all $y \in L^o \backslash Q$.

Then there is a compact connected subset $\hat{B}$ of L^o with the properties

(I) $\hat{B} \supset Q$,

(II) $\max \{f(x) \mid x \in \hat{B}\} = \max \{f(x) \mid x \in Q\}$.

Proof. Let Γ denote the class of all compact connected subsets of L^o containing Q. There is a sequence $\{B_i\}_{i \in \mathbb{N}}$ in Γ such that

$$\lim_{i \to \infty} \max \{f(x) \mid x \in B_i\} = \inf \{\max \{f(x) \mid x \in B\} \mid B \in \Gamma\} =: a.$$

For every $j \in \mathbb{N}$ $\hat{B}_j := \overline{\bigcup_{i \geq j} B_i}$ is an element of Γ. $\hat{B} := \bigcap_{j \in \mathbb{N}} \hat{B}_j$ is compact and connected as an intersection of a decreasing sequence of compact connected sets. In fact $\hat{B}$ is a member of Γ. It remains to show (II). From the definition of a we conclude that $\max \{f(x) \mid x \in \hat{B}\} \geq a$. Since $Q \subset B$ for all $B \in \Gamma$, we have

$$\alpha := \max \{f(x) \mid x \in Q\} \leq a.$$

For every $\varepsilon > o$ there is an index j such that $\max \{f(x) \mid x \in B_i\} \leq a + \varepsilon$ for all $i \geq j$. This implies $\max \{f(x) \mid x \in \hat{B}_j\} \leq a + \varepsilon$ and $\max \{f(x) \mid x \in \hat{B}\} \leq a + \varepsilon$. Consequently, $\max \{f(x) \mid x \in \hat{B}\} = a$.

To prove that $a=\alpha$, assume, to get a contradiction, that $a > \alpha$. Then $K := \{x \in \hat{B} \mid f(x)=a\}$ is a nonempty compact subset of $\hat{B}\backslash Q$ such that $f'(y) \neq 0$ for all $y \in K$.

Let U and T have the properties (I), (II), (III) of Prop.3 and set $\tilde{B} := T(\hat{B})$.

$\tilde{B}$ is compact and connected. $Q \subset G\backslash U$ implies $Q = T(Q) \subset \tilde{B}$. From the inequalities

$$f(T(x)) < f(x) \leq f(x^o) \text{ for all } x \in \hat{B} \cap U,$$

$$f(T(x)) = f(x) \leq f(x^o) \text{ for all } x \in \hat{B} \cap (G\backslash U)$$

we conclude that $\tilde{B} \subset L$. Since L^o is a connected component of L and $\tilde{B}$ a connected subset of L containing the nonempty subset Q of L^o we must even have $\tilde{B} \subset L^o$. Consequently, $\tilde{B} \in \Gamma$. But now, taking into account that

$$f(T(x)) < f(x) = a \text{ for all } x \in K,$$

$$f(T(x)) \leq f(x) < a \text{ for all } x \in \hat{B}\backslash K,$$

we arrive at the contradiction

$$a \leq \max \{f(y) \mid y \in \tilde{B}\} < a.$$

We are now ready to prove Lemma 2.
L^O is a connected component of L either by assumption or by Lemma 1. Q is a nonempty compact subset of L^O such that $f'(y) \neq o$ for all $y \in L^O\backslash Q$. Hence by Prop. 4 there must exist a

compact connected subset $\hat{B}$ of L^O with the properties

(I) $\hat{B} \supset Q$,

(II) $\max \{f(x) | x \in \hat{B}\} = \max \{f(x) | x \in Q\} = f(x^*)$.

But this yields $Q=\hat{B}$, stating the connectedness of Q. To show that Q is a connected component of $L(f(x^*))$, consider any connected subset O of $L(f(x^*))$ such that $Q \subset O$. From $O \subset L(f(x^o))$, $O \cap L^o \neq \emptyset$ and the fact that L^o is a connected component of $L(f(x^o))$ we conclude that $O \subset L^o$, hence $Q=O$ and Q must be a connected component of $L(f(x^*))$.

It is possible to improve Lemma 2, allowing certain isolated critical points of f in the set $\{x \in L^o | f(x) > f(x^*)\}$.

Recall that a point $x \in G$ is an isolated critical point of f iff $f'(x)=0$ and $f'(y) \neq 0$ for all y in a deleted neighbourhood of x. Every isolated critical point of f is either a point at which f has a strict relative extremum or a saddlepoint (a critical point x such that in every neighbourhood of x there are y_1, y_2 such that $f(y_1) < f(x) < f(y_2)$).

LEMMA 2a. Let L^o be a compact connected or path-connected component of $L := L(f(x^o))$ and let $x^* \in L^o$ be such that $f(x^*) < f(x^o)$.

Assume that all critical points of f in $L^o \backslash L(f(x^*))$ are isolated and relative minima of f. Then there is in fact no critical point in $L^o \backslash L(f(x^*))$ and therefore $L^o \cap L(f(x^*))$ is a connected component of $L(f(x^*))$.

Proof. Assume that $L^o \backslash L(f(x^*))$ contains critical points of f. Then for all sufficiently small $\lambda > f(x^*)$ the set of critical points of f contained in $\{x \in L^o | f(x) \geq \lambda\}$ is not empty and finite. Hence

$$\lambda^* := \max \{f(x) | x \in L^o \backslash L(f(x^*)) \text{ and } f'(x)=0\}$$

exists. $\lambda^* = f(x^o)$ would imply that there is a strict relative minimum x of f in L^o such that $f(x) = f(x^o)$. x would be an isolated point of the connected set L^o, yielding $L^o = \{x\}$, in contradiction to $x^* \in L^o$ and $f(x^*) < f(x)$. Hence $\lambda^* < f(x^o)$ and Lemma 2 shows that $L^o \cap L(\lambda^*)$ is a connected component of $L(\lambda^*)$. But now the existence of a strict relative minimum x of f such that $x \in L^o$ and $f(x) = \lambda^*$ implies $L^o \cap L(\lambda^*) = \{x\}$, again in contradiction to $x^* \in L^o$ and $f(x^*) < f(x)$.

We now use Lemma 2a to derive a condition assuring that f/L^o assumes its minimum in exactly one point.

LEMMA 3. Let L^o be a compact connected or path-connected component of $L := L(f(x^o))$ and assume that all critical points of f contained in L^o are isolated and relative minima of f.

Then there is exactly one critical point of f in L^o, the unique point at which f/L^o assumes its minimum.

Proof. The assertion is trivially true when L^o consists of only one point. Let us assume therefore that L^o contains more than one point. Then as a connected set, L^o contains infinitely many points and there must be noncritical points in L^o.

Hence there are also points x in L^o such that $f(x) < f(x^o)$. Let x^* be a (global) minimum of $f|L^o$. Then $f(x^*) < f(x^o)$ and Lemma 2a shows that there is no critical point of f in $L^o \setminus L(f(x^*))$ and that $L^o \cap L(f(x^*))$ is connected. Since every $x \in L^o \cap L(f(x^*))$ is a critical point of f, $L^o \cap L(f(x^*))$ is a finite connected set. Thus

$$L^o \cap L(f(x^*)) = \{x^*\}.$$

2. AN EXISTENCE AND UNIQUENESS THEOREM FOR SOLUTIONS OF NONLINEAR EQUATIONS

Lemma 3 has an interesting application to the problem of solving a nonlinear equation $F(x)=0$, when $F : G \to \mathbb{R}^n$ is a given C^1-mapping. For any smooth norm $\| . \|$ on $\mathbb{R}^n$ the following existence and uniqueness theorem can be obtained merely by stating Lemma 3 for the C^1-function

$$f : G \ni x \to \frac{1}{2} \| F(x) \|^2 \in \mathbb{R}.$$ [1]

[1] *We note that $\varphi : \mathbb{R}^n \ni y \to \frac{1}{2} \| y \|^2 \in \mathbb{R}$ is differentiable (even at $y=0$) and that every differentiable convex function defined on an open convex subset of $\mathbb{R}^n$ is also a C^1-function. (cf. [6] p. 173)*

Theorem 1. Let L^o be a compact connected or path-connected component of the set $S := \{x \in G \mid \|F(x)\| \leq \|F(x^o)\|\}$, where $\|.\|$ denotes a smooth norm on $\mathbb{R}^n$ and x^o a given point in G. Assume that

a) $F'(x)$ is regular whenever $x \in L^o$ and $F(x)=0$

b) $\|.\|'(F(x)) \circ F'(x) \neq 0$ whenever $x \in L^o$ and $F(x) \neq 0$.

Then there is exactly one $x^* \in L^o$ such that $F(x^*)=0$.

Proof. S is equal to the level set $L(f(x^o))$, where f is defined as above. From

$$f'(x) = \begin{cases} 0 & \text{whenever } F(x)=0 \\ \|F(x)\| \; \|.\|'(F(x)) \circ F'(x) & \text{whenever } F(x) \neq 0 \end{cases}$$

and the assumptions a) and b) we conclude that all critical points of f contained in L^o are isolated and relative minimum points of f. Thus by Lemma 3 there is exactly one critical point of f in L^o. But for every $x \in L^o$ the property of being a critical point of f is equivalent to the property of being a zero of F.

Remark. From $\| \, \|.\|'(F(x)) \, \| = 1$ whenever $F(x) \neq 0$ it follows that a) and b) are valid when $F'(x)$ is assumed to be regular for every $x \in L^o$. Essentially under this assumption Theorem 1 was stated without proof by P. Deuflhard in [2].

Theorem 1 can be used to improve convergence results for

iterative procedures for the solution of the equation $F(x)=0$, when they can be interpreted as descent methods for f. Before dealing with a concrete such method, we consider a consequence of Lemma 3 concerning more general descent methods.

3. CONVERGENCE PROPERTIES OF DESCENT METHODS

We restrict ourselves to descent methods for C^1-functions $f:G \to \mathbb{R}$ with the following properties: Given a starting point $x^0 \in G$, a sequence $x^1, x^2, \ldots$ is produced such that

(I) $(1-t)x^k + tx^{k+1} \in L(f(x^k))$ for all $t \in [0,1]$ and $k=0,1,\ldots$

(II) If the sequence remains in a compact connected subset K of G, there is a critical point $\hat{x}$ of f in K such that

$$\lim_{k\to\infty} f(x^k) = f(\hat{x}).$$

LEMMA 3a. Let L^0 be a compact connected or path-connected component of a level set $L(f(\bar{x}))$ and assume that all critical points of f contained in L^0 are isolated and relative minima of f. Then

a) There is exactly one relative minimum x^* of f in L^0.

b) For every $x^0 \in L^0$ and any descent method starting in x^0

and producing a sequence with the properties (I) and (II), the sequence converges to x^*.

Proof. The first part of the assertion is clear from Lemma 3. Given $x^o \in L^o$, any sequence with the properties (I) and (II) must remain in the path-connected component of $L(f(x^o))$ containing x^o. Therefore it is also contained in L^o. Lemma 3 shows that $\hat{x}=x^*$ and

$$f(\hat{x}) < f(x) \quad \text{for all } x \in L^o \setminus \{\hat{x}\}.$$

If the sequence would not converge to x^*, it would have a limit point $\tilde{x} \in L^o \setminus \{x^*\}$. But then we would arrive at the contradiction

$$f(\tilde{x}) = f(\hat{x}) < f(\tilde{x}).$$

In the case where f is given by

$$f: G \ni x \to \frac{1}{2} \|F(x)\|^2 \in \mathbb{R}$$

and the assumptions of Theorem 1 are valid, Lemma 3a can be replaced by the more convenient

LEMMA 3b. Under the assumptions of Theorem 1 let $\{x^k\}_{k \in \mathbb{N}_o}$ be a sequence in $L^o \setminus \{x^*\}$ such that $\{\| F(x^k) \|\}_{k \in \mathbb{N}_o}$ is decreasing.
Then $\lim_{k\to\infty} x^k = x^*$ is equivalent to each of the following two statements:

(1) $\lim_{k\to\infty} \| F(x^k) \| = 0$

(2) There is a sequence $\{\varepsilon_k\}_{k \in \mathbb{N}_0}$ of real nonnegative numbers with the properties

(a) $\|F(x^{k+1})\| \leq (1-\varepsilon_k) \| F(x^k) \|$ for every $k \in \mathbb{N}_0$

(b) $\sum_{k=0}^{\infty} \varepsilon_k = \infty$

Proof. Since x^* is the unique zero of $F|L^o$, clearly (1) implies that every subsequence of $\{x^k\}_{k \in \mathbb{N}_0}$ converges to x^*. The existence of a convergent subsequence of $\{x^k\}_{k \in \mathbb{N}_0}$ is assured by the compactness of L^o. The equivalence of (1) and (2) follows from the identities

$$\|F(x^{k+1})\| = \left(1 - \frac{\|F(x^k)\| - \|F(x^{k+1})\|}{\|F(x^k)\|}\right) \|F(x^k)\|$$

$$= \left\{ \prod_{i=0}^{k} \left(1 - \frac{\|F(x^i)\| - \|F(x^{i+1})\|}{\|F(x^i)\|}\right)\right\} \|F(x^o)\| ,$$

$$k = 0,1,\ldots$$

and the well known fact that for any sequence $\{\varepsilon_k\}_{k \in \mathbb{N}_0}$ in $[0,1[$ $\prod_{i=0}^{\infty} (1-\varepsilon_i)=0$ is equivalent to $\sum_{k=0}^{\infty} \varepsilon_k = \infty$.

As an illustrative application of Lemma 3b we will derive a convergence result for a damped Newton method applied first by A. M. Ostrowski to polynomial equations (comp. [9]) and recently by W. M. Häussler [5] to more general cases.

4. AN APPLICATION TO A DAMPED NEWTON METHOD

Let $F:G \to \mathbb{R}^n$ denote a C^1-mapping, $\|\cdot\|$ any norm on $\mathbb{R}^n$, x^o a given starting point in G and L^o the connected or path-connected component of

$$S := \{x \in G \mid \|F(x)\| \le \|F(x^o)\|\}$$

which contains x^o. Given $x,y \in \mathbb{R}^n$ and $\lambda \in [o,\infty[$, $[x,x-\lambda y]$ denotes the segment $\{x-ty \mid t \in [o,\lambda]\}$.

We first prove two lemmas extending slightly a result contained in [5] and somewhat more Exercise E.7.1-1. in [8].

LEMMA 4. Let $x \in L^o$ be such that $F'(x)$ is regular and $y := F'(x)^{-1}F(x) \neq 0$. Then there is a $\lambda > 0$ such that $[x,x-\lambda y] \subset L^o$ and moreover

$$\|F(x-ty)\| < \|F(x)\| \qquad \text{for all} \quad t \in]o,\lambda].$$

Proof. Choose $\lambda \in]0,1]$ such that $x-ty \in G$ and

$$\|F(x-ty)-F(x)-F'(x)(-ty)\| \le \frac{\|F(x)\|}{2\|y\|} \|ty\|$$

for all $t \in [0,\lambda]$. Then for any $t \in]0,\lambda]$ we have

$$\|F(x-ty)\| - (1-t)\|F(x)\| \le \|F(x-ty)-(1-t)F(x)\|$$
$$= \|F(x-ty)-F(x)-F'(x)(-ty)\| \le \frac{t}{2}\|F(x)\| ,$$

which yields $\| F(x-ty) \| \leq (1 - \frac{t}{2}) \| F(x) \| < \| F(x) \|$.

It follows that the path-connected set $[x,x-\lambda y]$ is contained in S and in fact in L^o, since $L^o \cap [x,x-\lambda y] \neq \emptyset$ and L^o is a connected or path-connected component of S.

LEMMA 5. Let x and y be as in Lemma 4. Assume that L^o is closed with respect to $\mathbb{R}^n$ and that we can find a constant $M > 0$ with the property

$$\| F'(x-ty) - F'(x) \| \leq Mt \| y \| \quad \text{whenever}$$

$[x,x-ty] \in L_x^o := \{x' \in L^o \,|\, \| F(x') \| \leq \| F(x) \| \}$ and $t \in [0,T]$, where $T \in \,]0,T(M)]$ and $T(M)$ is defined by

$$T(M) := \alpha(M) \min\{2, (1+4\alpha(M)^{-1})^{\frac{1}{2}} - 1\} \quad \text{and}$$

$$\alpha(M) := \| F(x) \| (M \| y \|^2)^{-1}.$$

Then $[x,x-Ty] \subset L_x^o$ and

$$\| F(x-ty) \| \leq ((2\alpha(M))^{-1} t^2 + |1-t|) \| F(x) \| < \| F(x) \|$$

for all $t \in \,]0,T[$.

Proof. The closedness of L^o with respect to $\mathbb{R}^n$ implies the closedness of L_x^o with respect to $\mathbb{R}^n$ and the latter the closedness of the interval

$$I := \{t \geq 0 \,|\, [x,x-ty] \subset L_x^o\}.$$

Lemma 4 shows that $I \neq \{0\}$.

For every $t \in I \cap [0,T]$ we have

$$\| F(x-ty) \| = \| F(x-ty)-F(x)-F'(x)(-ty)+(1-t)F(x) \|$$

$$\leq \| F(x-ty)-F(x)-F'(x)(-ty) \| + |1-t| \, \|F(x)\|$$

$$\leq \frac{M}{2} t^2 \|y\|^2 + |1-t| \, \|F(x)\| = ((2\alpha(M))^{-1}t^2 + |1-t|)\|F(x)\|$$

where $(2\alpha(M))^{-1}t^2+|1-t| < 1$ whenever $t \in]0,T(M)[$ and $(2\alpha(M))^{-1}T(M)^2+|1-T(M)| = 1$.

In order to prove the inclusion $[0,T] \subset I$, assume, to get a contradiction, that $I = [0,T^*]$ with $0 < T^* < T$. Then $\| F(x-T^*y)\| < \|F(x)\|$, and so there exists a $\tilde{T} > T^*$ such that $[x,x-\tilde{T}y]$ is contained in

$$S_x := \{x' \in G \mid \|F(x')\| \leq \|F(x)\| \} \subset S$$

and therefore in $S_x \cap L^o = L^o_x$, contradicting $\tilde{T} > T^*$.

Remarks.

1. It is not true in general that $\| F(x-ty) \| \leq \|F(x)\|$ for all $t \in [0,\alpha(M)]$. For a simple counterexample consider F defined on $\mathbb{R}$ by $F(t) := \frac{t^2}{2} + t - \frac{1}{2}$ together with $x=0$ and $M=1$. Then $y = -\frac{1}{2}$, $\alpha(M)=2$ and

$$|F(0-2y)| = |F(1)| = 1 > |F(0)| = \frac{1}{2}.$$

As a consequence of this, it seems to be impossible to prove the second assertion of E. 7. 1-1. in [8].

2. The function

$$a_M : [0,T(M)] \ni t \to (2\alpha(M))^{-1}t^2+|1-t| \in \mathbb{R}$$

has a unique minimum, given by

$$t_M := \min \{1,\alpha(M)\}.$$

In the case where $\alpha(M) \leq 1$, we have

$$a_M(t_M) = a_M(\alpha(M)) = 1 - \frac{\alpha(M)}{2} = 1 - \frac{t_M}{2} .$$

If $\alpha(M) \geq 1$, then

$$a_M(t_M) = a_M(1) = (2\alpha(M))^{-1} \leq \frac{1}{2} = 1 - \frac{t_M}{2} .$$

Hence $a_M(t_M) \leq 1 - \frac{t_M}{2}$ in every case.

For later purposes we state a corollary of Lemma 5 taking into account affine invariance as introduced in [3].

For a fixed $\tilde{x} \in G$ such that $F'(\tilde{x})$ is regular let $\tilde{L}^o$ denote the connected or path-connected component of

$$\tilde{S} := \{x \in G|\ \|F'(\tilde{x})^{-1}F(x)\| \leq \|F'(\tilde{x})^{-1}F(\tilde{x})\|\}$$

which contains $\tilde{x}$.

COROLLARY. Let $\tilde{L}^o$ be closed with respect to $\mathbb{R}^n$ and let $x \in \tilde{L}^o$ be such that $F'(x)$ is regular and $y := F'(x)^{-1}F'(x) \neq 0$. Assume that we can find a constant $\tilde{M} > 0$ with the property:

$\| F'(\tilde{x})^{-1}(F'(x-ty)-F'(x))\| \leq \tilde{M}t\|y\|$ whenever

$[x,x-ty] \subset \tilde{L}^o_x := \{x' \in \tilde{L}_o \,|\, \|F'(\tilde{x})^{-1}F(x')\| \leq \|F'(\tilde{x})^{-1}F(x)\|\}$ and

$t \in [0,\tilde{T}]$, where $\tilde{T} \in \,]0,\tilde{T}(\tilde{M})]$ with

$\tilde{T}(\tilde{M}) := \tilde{\alpha}(\tilde{M})\min\{2,(1+4\tilde{\alpha}(\tilde{M})^{-1})^{1/2} - 1\}$ and

$\tilde{\alpha}(\tilde{M}) := \|F'(\tilde{x})^{-1}F(x)\|(\tilde{M}\|y\|^2)^{-1}$.

Then $[x,x-\tilde{T}y] \subset \tilde{L}^o_x$ and

$\| F'(\tilde{x})^{-1}F(x-ty)\| \leq ((2\tilde{\alpha}(\tilde{M}))^{-1}t^2+|1-t|)\|F'(\tilde{x})^{-1}F(x)\|$

$< \|F'(\tilde{x})^{-1}F(x)\|$ for all $t \in \,]0,\tilde{T}[$.

Proof. Replace F in Lemma 5 by the mapping

$\tilde{F} : G \ni x \to F'(\tilde{x})^{-1}F(x) \in \mathbb{R}^n$.

Let us now describe a descent method for the solution of $F(x)=0$ motivated by the second remark following Lemma 5. Assume that L^o is closed with respect to $\mathbb{R}^n$ and that in the k-th step of an iteration ($k \in \mathbb{N}_0$) a point $x^k \in L^o$ and a constant $M_k > 0$ are produced such that

(I) $F'(x^k)$ is regular and $y^k := F'(x^k)^{-1}F(x^k) \neq 0$.

(II) $\|F'(x^k-ty^k)-F'(x^k)\| \leq M_k t\|y^k\|$ whenever

$[x^k,x^k-ty^k] \subset L^o_{x^k} := \{x \in L^o \,|\, \|F(x)\| \leq \|F(x^k)\|\}$ and

$t \in [0,t_k]$, where $t_k := \min\{1,\|F(x^k)\|(M_k\|y^k\|^2)^{-1}\}$.

Then, defining $x^{k+1} := x^k-t_k y^k$, we deduce from Lemma 5 and the second remark after it: $[x^k,x^{k+1}] \subset L^o_{x^k}$ and

$$\|F(x^{k+1})\| \leq (1 - \frac{t_k}{2})\,\|F(x^k)\| < \|F(x^k)\|.$$

We now have to consider three possible cases:

Case 1. $F'(x^{k+1})$ is regular and $y^{k+1} := F'(x^{k+1})^{-1}F(x^{k+1}) \neq 0$.

Case 2. $F'(x^{k+1})$ is singular.

Case 3. $F'(x^{k+1})$ is regular but $F(x^{k+1}) = 0$.

Assuming that we can find a suitable M_{k+1} in Case 1, we are able to continue the iteration. When Case 2 or 3 occurs we will stop. If the iteration does not stop after a finite number of steps, a sequence $\{x^k\}_{k \in \mathbb{N}_0}$ in L^0 is generated with $\{\|F(x^k)\|\}_{k \in \mathbb{N}_0}$ decreasing to a limit $\tau \geq 0$. If $\tau=0$, clearly every limit point of $\{x^k\}_{k \in \mathbb{N}_0}$ is a solution of $F(x)=0$.

If moreover L^0 is bounded and $F'(x)$ regular whenever $x \in L^0$ and $F(x)=0$, then $\{\|y^k\|\}_{k \in \mathbb{N}_0}$ and $\{\|x^{k+1}-x^k\|\}_{k \in \mathbb{N}_0}$ must tend to zero. But this implies that the set of limit points of $\{x^k\}_{k \in \mathbb{N}_0}$ is a connected subset of the discrete set of zeros of $F|L^0$ (cf. [9] Theorem 26.1). Hence $\{x^k\}_{k \in \mathbb{N}_0}$ converges to a zero of F. Finally if $\{M_k\}_{k \in \mathbb{N}_0}$ is bounded, the right-hand side of the inequality

$$\frac{\|F(x^k)\|}{M_k\|y_k\|^2} \geq \frac{1}{M_k \|F'(x^k)^{-1}\|^2\|F(x^k)\|}$$

is larger than 1 for all sufficiently large k. Hence t_k must be 1 when k is large. This means that $\{x^k\}_{k \in \mathbb{N}_0}$ is ultimately a Newton sequence.

Let us now consider the case $\tau > 0$. Then from

$$\tau < \|F(x^{k+1})\| \le (1 - \frac{t_k}{2}) \cdot \ldots \cdot (1 - \frac{t_o}{2})\|F(x^o)\|, \ k=0,1,\ldots,$$

we deduce $\lim_{k\to\infty} t_k=0$ and from this

$$\lim_{k\to\infty} M_k \|y^k\|^2 = \infty$$

If it is known that $\{M_k\}_{k \in \mathbb{N}_o}$ is bounded, F' must be singular at every limit point x of $\{x^k\}_{k \in \mathbb{N}_o}$. If moreover L^o is bounded and the M_k are bounded below by a positive constant M, we have

$$\|x^{k+1}-x^k\| = t_k\|y^k\| \le \|F(x^k)\| (M\|y^k\|)^{-1}$$

for all sufficiently large k, showing

$$\lim_{k\to\infty} \|x^{k+1}-x^k\| = 0.$$

This implies the convergence of $\{x^k\}_{k \in \mathbb{N}_o}$ when the set of singular points of $F \mid L^o$ is finite.

Combining these results with Theorem 1 we obtain the following convergence statement:

THEOREM 2. Let $\|\cdot\|$ be a smooth norm on $\mathbb{R}^n$ and $x^o \in G$ be such that the connected or path-connected component L^o of

$$S := \{x \in G \mid \|F(x)\| \le \|F(x^o)\|\}$$

containing x^o is compact. Assume that

a) $F'(x)$ is regular whenever $x \in L^o$.

b) A constant $M > 0$ is known such that

$$\| F'(w) - F'(x) \| \leq M \| w - x \| \qquad \text{for all } w, x \in L^o.$$

Then we can state:

(I) There is exactly one $x^* \in L^o$ such that $F(x^*) = 0$.

(II) If for every $x \in L^o \setminus \{x^*\}$ we set

$t(x) := \min\{1, \|F(x)\| (M \|F'(x)^{-1} F(x)\|^2)^{-1}\}$ and

$s(x) := x - t(x) F'(x)^{-1} F(x)$, then $[x, s(x)]$ is contained in L^o and we have

$$\| F(s(x)) \| \leq \left(1 - \frac{t(x)}{2}\right) \| F(x) \|.$$

(III) The sequence defined by

$$x^{k+1} := \begin{cases} s(x^k) & \text{when } x^k \neq x^*, \\ x^* & \text{otherwise} \end{cases} \qquad k = 0, 1, \ldots.$$

converges to x^*. If $x^k \neq x^*$ for every $k \in \mathbb{N}_0$, then $t(x^k) = 1$ for all sufficiently large k, i.e. $\{x^k\}_{k \in \mathbb{N}_0}$ is ultimately a Newton sequence.

Remark. Theorem 2 is not affine invariant in the sense of [3]. Nevertheless it can be converted into an affine invariant convergence theorem simply by applying it to $F'(x^o)^{-1} F$ instead of F.

5. A CONTINUATION METHOD

In [7] D. Leder has developed a continuation method from a descent method due to Goldstein and Armijo. Using the same idea, a continuation method can also be derived from the damped Newton method studied in the last section.

In what follows, let I denote an open interval containing $[0,1]$, $F : G \to \mathbb{R}^n$ and $H : G \times I \to \mathbb{R}^n$ C^1-mappings and x^0 a given starting point in G such that

$$H(\cdot,1) = F \text{ and } H(x^0,0) = 0.$$

For a fixed real $m \neq 0$, we set

$$\tilde{F} : G \times I \ni (x,t) \to (H(x,t), m(1-t)) \in \mathbb{R}^n \times \mathbb{R}.$$

(x^*, t^*) is a zero of $\tilde{F}$ iff $t^*=1$ and x^* is a solution of the equation $F(x)=0$. Hence, in order to solve the equation $F(x)=0$, it might be useful to study the case where Theorem 2 applies to $\tilde{F}$ but not necessarily to F itself. As our last theorem will show, this study leads to a convergence statement about a continuation method with variable steplengths. Let us first collect the main quantities required for the application of Theorem 2 to $\tilde{F}$.

The derivative of $\tilde{F}$ at $(x,t) \in G \times I$ is given by the mapping

$$\tilde{F}'(x,t) : \mathbb{R}^n \times \mathbb{R} \ni (h,\tau) \to (D_1H(x,t)(h)+\tau D_2H(x,t)(1), -m\tau) \in \mathbb{R}^n \times \mathbb{R}.$$

$\tilde{F}'(x,t)$ is regular if $D_1H(x,t)$ is regular, in which case $\tilde{F}'(x,t)^{-1}(h,\tau) = (D_1H(x,t)^{-1}(h+\frac{\tau}{m} D_2H(x,t)(1)),-\frac{1}{m}\tau)$ for all $(h,\tau) \in \mathbb{R}^n \times \mathbb{R}$. This yields

$\tilde{F}'(x,t)^{-1}\tilde{F}(x,t) = (D_1H(x,t)^{-1}(H(x,t)+(1-t)D_2H(x,t)(1)),t-1)$

whenever $\tilde{F}'(x,t)$ is regular.

Stating now Theorem 2 for $\tilde{F}$ we get

THEOREM 3. Let $\|\,.\,\|$ be a smooth norm on $\mathbb{R}^n \times \mathbb{R}$ and assume that

a) The connected or path-connected component $\tilde{L}^o$ of $\tilde{S} := \{(x,t) \in G \times I \mid \|\tilde{F}(x,t)\| \leq \|(0,m)\|\}$ containing $(x^o,0)$ is compact.

b) $D_1H(x,t)$ is regular whenever $(x,t) \in \tilde{L}^o$.

c) A constant $\tilde{M} > 0$ is known such that $\|\tilde{F}'(x_1,t_1)-\tilde{F}'(x_2,t_2)\| \leq \tilde{M}\|(x_1-x_2,t_1-t_2)\|$ for all (x_1,t_1), $(x_2,t_2) \in \tilde{L}^o$.

Then

(I) There is exactly one $x^* \in G$ such that $(x^*,1) \in \tilde{L}^o$ and $F(x^*)=0$.

(II) If for every $(x,t) \in \tilde{L}^o \setminus \{(x^*,1)\}$ we set

$\tau(x,t) := \min\,\{1,\|\tilde{F}(x,t)\|\,(\tilde{M}\|\tilde{F}'(x,t)^{-1}\tilde{F}(x,t)\|^2)^{-1}\}$,

$\tilde{s}(x,t) := x-\tau(x,t)D_1H(x,t)^{-1}(H(x,t)+(1-t)D_2H(x,t)(1))$ and

$\tilde{\sigma}(x,t) := t-\tau(x,t)(t-1) = (1-\tau(x,t))t + \tau(x,t)\cdot 1$, then

$[(x,t),(\tilde{s}(x,t),\tilde{\sigma}(x,t))]$ is contained in $\tilde{L}^o$ and we have

$$\|\tilde{F}(\tilde{s}(x,t),\tilde{\sigma}(x,t))\| \leq (1-\frac{\tau(x,t)}{2})\,\|\tilde{F}(x,t)\|.$$

(III) The sequence defined by $(x^o,t_o) := (x^o,0)$ and

$$(x^{k+1},t_{k+1}) := \begin{cases} (\tilde{s}(x^k,t_k),\tilde{\sigma}(x^k,t_k)) & \text{when } (x^k,t^k)\neq(x^*,1) \\ (x^*,1) & \text{otherwise} \end{cases}$$

for all $k=0,1,..$ converges to $(x^*,1)$. If $(x^k,t_k)\neq(x^*,1)$ for every $k \in \mathbb{N}_o$, then $\tau(x^k,t_k) = 1 = t_{k+1}$ for all sufficiently large k and $t_k < t_{k+1}$ whenever $t_k < 1$.

Remark. Recalling the remark after Theorem 2, we observe that an application of Theorem 3 to $\tilde{F}'(x^o,0)^{-1}\tilde{F}$ instead of $\tilde{F}$ leads to an affine invariant convergence theorem.

ACKNOWLEDGMENT

The author wishes to thank Dr. G. Berthiaume for his careful reading of the manuscript.

REFERENCES

[1] Berger, M. S., Berger, M. S., Perspectives in Nonlinearity, an Introduction to Nonlinear Analysis, W. A. Benjamin, Inc., New York and Amsterdam (1968).

[2] Deuflhard, P., A Modified Newton Method for the Solution of Ill-Conditioned Systems of Nonlinear Equations with Application to Multiple Shooting, Numer. Math., 22, (1974) 289-315.

[3] Deuflhard, P., Heindl, G., Affine Invariant Convergence Theorems for Newton's Method and Extensions to Related Methods, Technische Universität München, Institut für Mathematik TUM-MATH-7723 (1977).

[4] Dugundji, J., Topology, Allyn and Bacon, Inc., Boston (1972).

[5] Häussler, W. M., Zum gedämpften Newton-Verfahren, TU Braunschweig, Institut für Rechentechnik (1976).

[6] Holmes, R. B., Geometric Functional Analysis and its Applications, Springer, New York, Heidelberg and Berlin (1975).

[7] Leder, D., Automatische Schrittweitensteuerung bei global konvergenten Einbettungsmethoden, ZAMM 54 (1974), 319-324.

[8] Ortega, J. M., Rheinboldt, W. C., Iterative Solution of Nonlinear Equations in Several Variables, Acad. Press, New York and London (1970).

[9] Ostrowski, A. M., Solution of Equations in Euclidean and Banach Spaces, Acad. Press, New York and London (1973).

EXISTENCE AND DISCRETE APPROXIMATION OF BIFURCATION POINTS

H. Jeggle

Technische Universität Berlin
Berlin, Federal Republic of Germany

I. INTRODUCTION

In connection with the application of homotopy methods as well as with stability problems as the Navier-Stokes equations for steady motions of an incompressible viscous fluid or biological and chemical reaction models there arises the question of the existence of bifurcating solutions of the respective equations. Hence, the aim of this paper is to discuss the existence of bifurcation points, i.e., parameter values $\lambda_0 \in \mathbb{R}$ such that the problem

(A) $\lambda Bx - Ax = 0$

has solutions $(\lambda, x(\lambda))$ where λ is close to λ_0 and $x(\lambda)$ has

ISBN 0-12-729250-0

arbitrarily small but nonzero norm. The operators A,B are supposed to be nonlinear mappings between real Banach spaces E,F with AO = O and BO = O.

The work of Krasnoselski [9] on this problem marked the appropriate method. He had the identity for A and supposed B to be compact. Petryshyn [11] succeeded in extending the results of [9] to the case where A is A-proper and B compact. He gave there also a survey on related results. In this paper we undertake a further generalization introducing the very general constructive concepts of discrete approximations due to Stummel [12] and the approximation-regular operators [2,5,6]. Doing that we can at once dispense with the construction of complicated approximation schemes and demonstrate the existence results for that extensive class of operators what leads moreover directly to the convergence of an approximation process for the bifurcation points.

The following two sections contain the precise formulation of the problem and introduce the fundamentals concerning the discrete approximation of spaces and the class of admissible operators including several examples. Then, following Krasnoselki [9] the problem (A) is linearized. Assuming certain properties of Fredholm type for the linearized operators there is shown that at most the eigenvalues of the linearized problem can be bifurcation points. On the other hand at least those among the eigenvalues which have odd algebraic multiplicity possess the desired property. The proof of this result rests heavily on a combination of approximate solvability results for linear approximation-regular operator equations and properties of the Brouwer degree of linear mappings. Having approached this theorem, it is nearby to use the convergence theory for linear eigenvalue problems for the discussion of the approximate calculation of bifurcation points, what is indicated in the last section.

Summarizing our method consists actually in a demonstration how powerful a linearization can be.

2. PRELIMINARIES

Let E,F be real Banach spaces. We consider the eigenvalue problem

(1) $C(\lambda)u \equiv \lambda Bu - Au = 0, \quad u \in D(C) \subset E,$

with, in general, nonlinear operators $A : D(C) \to F$, $B : D(C) \to F$. These operators are supposed to have the following properties:

(2) $A0 = 0, \; B0 = 0$

(3) A and B have the Fréchet derivatives A' and B' respectively at 0.

A point $\lambda_0 \in \mathbb{R}$ is called a *bifurcation point of* (A,B), if to each sufficiently small $\delta>0$ there corresponds a $\rho \equiv \rho(\delta)>0$ such that, for any $r \in (0,\rho)$, the eigenvalue problem

(1) has a solution $(\lambda, x(\lambda))$ with $|\lambda-\lambda_0|<\delta$ and $0<\|x(\lambda)\| = r$. The first aim consists in establishing conditions for the existence of such bifurcation points of (A,B). For this purpose we introduce the linearized eigenvalue problem.

(4) $C'(\lambda)u \equiv \lambda B'u - A'u = 0, \quad u \in E.$

A point $\lambda_0 \in \mathbb{R}$ is called an *eigenvalue of* (A',B'), if there exists a solution $(\lambda_0, x(\lambda_0)) \neq (0,0)$ of (4). The corresponding element $x(\lambda_0)$ is called an *eigenvector* or *principal vector of first grade* of (A',B') to λ_0. For $n\geq 2$, $y \neq 0$ is a principal vector of grade n, if there exists a principal vector x of grade n-1 such that there holds $\lambda B'y - A'y = B'x$. The dimensions of the linear hulls of all eigenvalues and all principal vectors of (A',B') to λ_0 are called the *geometric* and *algebraic multiplicities of* λ_0 respectively.

For the approximate evaluation of the bifurcation points, there is used a sequence of eigenvalue problems in real Banach spaces E_h, F_h

(5) $C_h(\lambda)u_h \equiv \lambda B_h u_h - A_h u_h = 0, \; u_h \in D(C_h) \subset E_h, \; h \in H_0$

where H_0 denotes a sequence of indices. $A_h : D(C_h) \to F_h$ and $B_h : D(C_h) \to F_h$ are, in general, nonlinear operators with the properties

(6) $A_h 0 = 0, \; B_h 0 = 0, \; h \in H_0.$

(7) For each $h \in H_0$, A_h and B_h have the Fréchet derivatives A'_h and B'_h respectively at 0.

Subsequently, $H, H', \ldots$ and $H_1, H_2, \ldots$ denote subsequences and final pieces of H respectively, not necessarily always the same ones. Since there can not be any confusions, a notational distinction between the zero elements or the norms of E, F, E_h, F_h will not be made.

The condition (7) justifies the linearization of the approximate problems,

$$(8) \qquad C'_h(\lambda)u_h = \lambda B'_h u_h - A'_h u_h = 0, \; u_h \in E_h, \; h \in H_0.$$

In order to create a relation between the problems (1) and (4) and their approximations (5) and (8) respectively as we know that from differential or integral equations and their discretizations we have to impose certain conditions on the spaces and the operators. As for the spaces, we assume in the following that they form a special discrete approximation in the sense of Stummel [12], i.e., that in the case of E there is given a sequence of restriction operators $R_h^E : E \to E_h$ with

$$(9) \qquad \|R_h u\| \to \|u\| \, (h \in H) \quad \forall u \in E$$

$$(10) \qquad \|R_h(\alpha u + \beta v) - \alpha R_h u - \beta R_h v\| \to 0 \, (h \in H), \forall \alpha, \beta \in \mathbb{R} \text{ and } u, v \in H.$$

A similar restriction (R_h^F) is supposed to exist for F, F_h. Due to this assumption we can use the notion of a discrete *convergence* of a sequence $u_h \in E_h$ (or F_h) to the limit $u \in E$ (or F) introduced by Stummel [12],

$$u_h \to u \, (h \in H) \text{ if } \|u_h - R_h u\| \to 0 \, (h \in H).$$

This convergence has the main properties of the usual norm convergence in Banach spaces. For operators T, T_h, the convergence is defined subsequently by

$$T_h \to T(h \in H) \text{ if } u_h \to u(h \in H) \Rightarrow T_h u_h \to Tu(h \in H)$$

for all $u \in D(T)$ and all $u_h \in D(T_h)$. In a general discrete approximation a weak convergence cannot be defined [2]. However, in the case of reflexive spaces we can consider discrete approximations of dual pairs $\langle E^*, E\rangle$. Assuming that E, E_h and E^*, E_h^* are discrete approximations in the sense introduced above, a condition of compatibility is added: For all $u \in E$, $\varphi \in E^*$, $u_h \in E_h$ and $\varphi_h \in E_h^*$ with $u_h \to u$, $\varphi_h \to \varphi$ $(h \in H)$ there holds $\langle \varphi_h, u_h\rangle \to \langle \varphi, u\rangle$ $(h \in H)$. In such a construction the *weak convergence* of a sequence $u_h \in E_h$ to a limit $u \in E$ is defined by

$$u_h \rightharpoonup u(h \in H) \text{ if } \langle \varphi_h, u_h\rangle \to \langle \varphi, u\rangle \quad (h \in H)$$

for all $\varphi \in E^*$ and all $\varphi_h \in E_h^*$ with $\varphi_h \to \varphi$ $(h \in H)$.

This weak convergence again possesses all important properties known from the usual case. In particular, there holds the well-known weak compactness property.

(11) Let E, E_h be reflexive and separable real Banach spaces such that $\langle E^*, E\rangle$, $\langle E_h^*, E_h\rangle$ form a compatible discrete approximation. Any bounded sequence $u_h \in E_h$ contains then a weakly convergent subsequence.

For the details we refer to [7] and the references quoted there.

We will use the following notation: K(a,r), B(a,r) and S(a,r) are the open or closed ball and the sphere with center at a and radius r in E or F. Similarly we use $K_h(a,r)$, $B_h(a,r)$ and $S_h(a,r)$ for the respective sets in E_h or F_h.

3. APPROXIMATION-REGULAR OPERATORS

There is not yet introduced the above mentioned relation between the operators C and C_h. However, in a series of recent papers (cf. [2,3,4,5,6,8,13]) devoted to constructive existence results and the theory of approximation methods for inhomogeneous equations as well as eigenvalue problems there was demonstrated that the class of the so-called approximation-regular pairs of operators plays a key-role in connection with such investigations for linear and nonlinear mappings. This concept is used here again.

A pair $T:D(T) \subset E \to F$, $T_h:D(T_h) \subset E_h \to F_h$ is called *approximation-regular* (*a-regular*) if for any subsequence $H \subset H_0$ and any bounded sequence $u_h \in D(T_h)$, $h \in H$, the convergence

$$T_h u_h \to v \in F (h \in H)$$

implies that there exists an element $u \in D(T)$ and a subsequence $H' \subset H$ such that

$$u_h \to u(h \in H'),\ Tu = v.$$

In view of the various papers on this subject (cf. the papers cited above and their references) we must neither comment this definition nor repeat many examples of the numerous criteria for a-regularity. However, since they are particularly important in connection with the pairs of problems 2.(1), 2.(5) and 2.(4), 2(8), we present some examples for a-regular sums of operators. Unfortunately, the sums of a-regular operators are in general not a-regular [10].

Discussing convergence properties of approximate problems the discretely compact operator sequences play a remarkable role similar to that of the a-regular operators (cf. [4,6,12]). A sequence $T_h : D(T_h) \to F_h$ is called *discretely compact* if for any subsequence $H \subset H_0$ and any bounded sequence $u_h \in D(T_h)$ there exists a subsequence $H' \subset H$ and an element $y \in F$ such that $T_h u_h \to y(h \in H')$.

(1) Let T, T_h be an a-regular pair and assume that (K_h) is discretely compact with $K_h \to K(h \in H)$. Then, $T+K, T_h+K_h$ is a-regular.

Proof: Let $u_h \in D(T_h+K_h)$ be bounded with $T_h u+K_h u \to w(h \in H)$. Since (K_h) is discretely compact we have a subsequence $H' \subset H$ and $v \in F$ such that $K_h u_h \to v(h \in H')$. Then, $T_h u_h \to w-v(h \in H')$

follows immediately. From the a-regularity of T, T_h we deduce the existence of $u \in D(T+K)$ and $H'' \subset H'$ such that $u_h \to u(h \in H'')$ and $Tu = w-v$. The convergence $K_h \to K$ and the uniqueness of limits imply $Ku = v$. Hence

$$u_h \to u(h \in H''), \quad Tu + Ku = w.$$

Since the identities I, I_h form a trivial pair of a-regular operators, we observe that $I+K$, I_h+K_h are a-regular if (K_h) satisfies the assumptions of (1). The following theorems model situations met occasionally in connection with partial differential operators and suitable approximations.

(2) Let $\langle E^*, E\rangle$, $\langle E_h^*, E_h\rangle$ be a compatible discrete approximation of a reflexive and separable dual pair E^*, E. Assume that $A: D \to E^*$, $A_h: D_h \to E_h^*$ are defined with $T: D\times D \to E^*$ and $T_h: D_h\times D_h \to E_h^*$ by $A(u) \equiv T(u,u)$ and $A_h(u_h) \equiv T_h(u_h,u_h)$.

Let the following conditions hold:

(i) For any sequence $u_h \in D_h$ with $u_h \rightharpoonup u(h \in H)$ there holds $u \in D$.

(ii) For any $u \in D$ there exists $(T(u,.))^{-1}$ on E^*

(iii) $(T_h(u_h,.))$ is uniformly strictly monotone for any $u_h \in D_h$, i.e., there exists $\gamma > 0$ such that

$$\gamma \| x_h - y_h \|^2 \le \langle T_h(u_h,x_h) - T_h(u_h,y_h),\ x_h - y_h\rangle$$

for all $x_h, y_h \in D_h$ and any $u_h \in D_h$, $h \in H_0$.

(iv) <u>For any sequences</u> $u_h, y_h \in D_h$ <u>and elements</u> $u, y \in D$ <u>there follows from</u> $u_h \rightharpoonup u (h \in H)$, $y_h \to y (h \in H)$ <u>that there converges</u> $T_h(u_h, y_h) \to T(u,y) (h \in H)$.

<u>Then</u> A, A_h <u>is a-regular</u>.

Proof: Let $u_h \in D_h$ be a bounded sequence with $A_h u_h \to v \in E^*$ $(h \in H)$. Hence, by 2.(11) and (i) there exist $H' \subset H$ and $u \in D$ such that $u_h \rightharpoonup u (h \in H')$. Assume that $y_h \in D_h$ is an arbitrary convergent sequence with limit y. Then we take from (iii) for all $h \in H_0$

$$\gamma \| u_h - y_h \|^2 \leq \langle T_h(u_h, u_h) - T_h(u_h, y_h), u_h - y_h \rangle$$

and

$$\gamma \| u_h - y_h \|^2 \leq \langle A_h u_h - T_h(u_h, y_h), u_h - y_h \rangle.$$

The limit process $h \in H'$ delivers now with (iv) and the properties of weak convergence

$$\gamma \lim\sup \| u_h - y_h \|^2 \leq \langle v - T(u,y), u - y \rangle.$$

Using (ii) we conclude that there exists a unique $y_0 \in D$ such that $T(u, y_0) = v$, what means that $u = y_0$ and $A(u) = v$.

We have yet to show the convergence $u_h \to u$.

From (iii) follows

$$\gamma \| u_h - R_h u \|^2 \leq \langle T_h(u_h, u_h) - T_h(u_h, R_h u), u_h - R_h u \rangle.$$

Due to (iv), we get again by a limit process that

$$\|u_h - R_h u\| \to 0 (h \in H').$$

As a corollary, we have the a-regularity of $A+K$, A_h+K_h, if A, A_h and K, K_h satisfy the conditions of (2) and (1) respectively.

(3) Let $\langle E^*, E\rangle$, $\langle E_h^*, E_h\rangle$ be a compatible discrete approximation of a reflexive and separable dual pair E^*, E.
Assume that $A: D \to E^*$, $A_h: D_h \to E_h^*$ are defined with $T: D \times D \to E^*$ and $T_h: D_h \times D_h \to E_h^*$ by $A(u) \equiv T(u,u)$ and $A_h(u_h) \equiv T_h(u_h, u_h)$.
Let for A, A_h and $B: D \to E^*$, $B_h: D_h \to E^*$ the following conditions hold:

(i) For any sequence $u_h \in D_h$ with $u_h \rightharpoonup u$ $(h \in H)$ there holds $u \in D$.

(ii) For any $u \in D$ there exists $(T(u,.))^{-1}$ on E^*.

(iii) $(T_h(u_h,.))$ is uniformly monotone for any $u_h \in D_h$, i.e.

$$0 \le \langle T_h(u_h, x_h - T_h(u_h, y_h), x_h - y_h\rangle$$

for all $x_h, y_h \in D_h$ and any $u_h \in D_h$, $h \in H_0$.

(iv) For any sequences $u_h, y_h \in D_h$ and elements $u, y \in D$ there follows from $u_h \rightharpoonup u (h \in H)$ and $y_h \to y (h \in H)$ that there converges $T_h(u_h, y_h) \to T(u,y) (h \in H)$.

(v) (B_h) maps bounded sequences into bounded sequences and B, B_h are uniformly of type (M), i.e., for any sequence $u_h \in D_h$ there follows from

$$u_h \rightharpoonup u,\ B_h u_h \rightharpoonup z,\quad \limsup \langle B_h u_h, u_h\rangle \le \langle z, u\rangle$$

that $Bu = z$.

(vi) For any sequence $u_h \in D_h$ there follows from

$$u_h \rightharpoonup u, \ \limsup \langle B_h u_h, u_h - R_h u\rangle \le 0$$

that $u_h \to u$.

Then, A+B, $A_h + B_h$ is a-regular.

Proof: Let $u_h \in D_h$ be a bounded sequence with $A_h u_h + B_h u_h \to v (h \in H)$. By 2.(11) and (i) we have $u \in D$ and $H' \subset H$ such that $u_h \rightharpoonup u (h \in H')$. Let $y_h \in D_h$ by any arbitrary convergent sequence with limit y. Defining

$$\varphi_h \equiv \langle T_h(u_h, u_h) - T_h(u_h, y_h) + B_h u_h, u_h - y_h\rangle$$

we conclude with (iv) by a limit process that

$$(4) \qquad \varphi_h \to \langle v - T(u,y), u - y\rangle \ (h \in H').$$

On the other side follows from (ii)

$$\varphi_h \ge \langle B_h u_h, u_h - y_h\rangle, \ h \in H'.$$

Hence,

$$\limsup \varphi_h \ge \limsup \langle B_h u_h, u_h - y_h\rangle$$

and from (4)

$$\limsup \langle B_h u_h, u_h - y_h\rangle \le \langle v - T(u,y), u - y\rangle.$$

Since (u_h) and hence $(B_h u_h)$ are bounded there exists a subsequence $H'' \subset H$ and a $z \in E^*$ such that

(5) $B_h u_h \rightharpoonup z \, (h \in H'')$,

Thus we get

$$\lim\sup \langle B_h u_h, u_h - y_h\rangle - \langle z, u-y\rangle \le \langle v-z-T(u,y), u-y\rangle.$$

Due to (ii) we have a uniquely determined $y_o \in D$ such that $T(u,y_o) = v-z$. Hence,

$$\lim\sup \langle B_h u_h, u_h - R_h y_o\rangle \le \langle z, u-y_o\rangle.$$

Using (5) there follows

(6) $\lim\sup \langle B_h u_h, u_h\rangle \le \langle z, u\rangle$,

that leads with (v) to $Bu = z$. Rewriting the preceding inequality we remark

$$\lim\sup \langle B_h u_h, u_h\rangle \le \langle Bu, u\rangle.$$

The weak convergence $B_h u_h \rightharpoonup z \, (h \in H'')$ delivers

$$\lim\sup \langle B_h u_h, u_h - R_h u\rangle = \lim\sup \langle B_h u_h, u_h\rangle - \langle z, u\rangle.$$

Observing (6) this implies

$$\lim\sup \langle B_h u_h, u_h - R_h u\rangle \le 0.$$

Due to (vi) we have now $u_h \to u(h \in H")$. Using (iv) there follows $T_h(u_h, u_h) \to Au(h \in H")$ and, subsequently, $B_h u_h \to v - Au(h \in H")$. This includes $\langle B_h u_h, u_h \rangle \to \langle v-Au, u \rangle$. Thus, by (v) with $z \equiv v-Au$ we derive $Bu = v-Au$. Eventually, we have shown $u_h \to u(h \in H")$ and $Au + Bu = v$.

For the sake of convenience we add some results that will be useful for our further considerations. A pair T, T_h is called *consistent*, if there exists for each $u \in D(T)$ a sequence $u_h \in D(T_h)$ such that $u_h \to u$, $T_h u_h \to Tu(h \in H)$. Grigorieff [3] has shown:

(7) Let E, E_h and F, F_h be discrete approximations with F separable. Assume that (T_h) is discretely compact and consistent with T. Then T is compact.

The following theorem [2] will prove to be of basic importance, it is valid for bounded linear mappings, $B(E,F)$ denotes the Banach space of these operators from E into F.

(8) The following two statements are equivalent.

(i) The operator $T \in B(E,F)$ is injective, the sequence $T_h \in B(E_h, F_h)$ is a-regular with T and there exists a final piece $H_1 \subset H$ such that $\ker(T_h) = \{0\}$ implies $\operatorname{im}(T_h) = F_h$, $h \in H_1$.

(ii) T^{-1} exists in $B(F,E)$, for a final piece $H_1 \subset H$ there exists T_h^{-1} in $B(F_h, E_h)$ and there holds the convergence

$$T_h^{-1} \to T^{-1} (h \in H_1).$$

In both cases the pair T, T_h is consistent.

The following corollary contains a slight modification.

(9) Let there exist a final piece $H_1 \subset H$ such that for $T_h \in B(E_h, F_h)$ there holds $\ker(T_h) = \{0\}$ implies $\mathrm{im}(T_h) = F_h$, $h \in H_1$. Then the following two statements are equivalent.

(i) The operator $T \in B(E,F)$ is injective and a-regular with the sequence $T_h \in B(E_h, F_h)$.

(ii) The operator T is surjective, the sequence (T_h) is consistent with T and there exist a $\gamma > 0$ and a final piece $H_1 \subset H$ such that there holds

$$\gamma \|u_h\| \leq \|T_h u_h\|, \quad u_h \in E_h, \quad h \in H_1.$$

4. EXISTENCE OF BIFURCATION POINTS

Following the patterns set by Krasnoselski [9] and Petryshyn [11] we subsequently use a linearization principle which permits to introduce results from degree theory for linear operators.

In order to verify the assumptions required for defining the Brouwer degree for the approximating mappings the discrete approximations must be specialized. For the remaining part of the paper we assume that E_h, F_h are approximately oriented real Banach spaces of equal finite dimension, $h \in H_o$. Recall that this assumption implies

(1) $\ker(T_h) = \{0\} \Rightarrow \operatorname{im}(T_h) = F_h,\ T \in B(E_h, F_h),\ h \in H_o.$

Moreover we will need the following properties from E, E_h:

(2) $u_h \in S_h(0,r) : u_h \to u (h \in H) \Rightarrow u \in S(0,r).$

The first theorem excludes all values of the parameter λ from the set of bifurcation points of (A,B) but the eigenvalues of (A',B').

(3) <u>Let</u> $G \subset \mathbb{R}$ <u>be an open set such that</u>

$$\ker(C'(\lambda)) = \{0\} \Rightarrow \operatorname{im}(C'(\lambda)) = F,\ \lambda \in G.$$

<u>If</u> $\lambda_o \in G$ <u>is not an eigenvalue of</u> (A',B'), <u>then</u> λ_o <u>is not a bifurcation point of</u> (A,B).

Proof: If λ_o is not an eigenvalue of (A',B'), then $C(\lambda_o)$ is injective. Due to the assumption, we have $C'(\lambda_o)^{-1} \in B(F,E)$. Take the interval

(4) $|\lambda - \lambda_o| < \alpha \equiv (4\,\|B'\|\ \|C_o'(\lambda_o)^{-1}\|)^{-1}.$

Observing 2.(3) it is possible to find a ball $K(0,\rho)$ such that there holds

(5) $\|Ax - A'x\| \leq (4\|C'(\lambda_o)^{-1}\|)^{-1}\|x\|$

and

(6) $\|Bx - B'x\| \leq (4\|C'(\lambda_o)^{-1}\|(|\lambda_o| + \alpha))^{-1}\|x\|$

for all $x \in K(O,\rho)$. Suppose now that $(\lambda, x(\lambda))$ is a solution of 2.(1) with $|\lambda - \lambda_o| < \alpha$ and $x(\lambda) \in K(O,\rho)$. Then we have

$$\begin{aligned}\|x(\lambda)\| &= \|C'(\lambda_o)^{-1}(C'(\lambda_o)x(\lambda))\| \\ &\leq \|C'(\lambda_o)^{-1}\| \, \|(\lambda_o - \lambda)B'x(\lambda) \\ &\quad + \lambda(B'x(\lambda) - Bx(\lambda)) + C(\lambda)x(\lambda) \\ &\quad + (Ax(\lambda) - A'x(\lambda))\|\end{aligned}$$

Using (4), (5) and (6) there follows

$$\|x(\lambda)\| \leq \frac{3}{4} \|x(\lambda)\| .$$

Hence $x(\lambda) = O$, what means that λ_o cannot be a bifurcation point.

The assumption of (3) can be made accessible by a variety of conditions, we must only recall certain Fredholm properties. Fortunately, we can deduce it from the a-regularity too.

(7) _Let_ $G \subset \mathbb{R}$ _be an open set such that_ $C'(\lambda)$, $C_h'(\lambda)$ _is a-regular for each_ $\lambda \in G$. _If_ $\lambda_o \in G$ _is not an eigenvalue of_ (A',B'), _then_ λ_o _is not a bifurcation point of_ (A,B).

Proof: If λ_o is not an eigenvalue of (A',B'), then $C'(\lambda_o)$ is injective. With the a-regularity of $C'(\lambda_o)$, $C_h'(\lambda_o)$ and (1) we take from 3.(8) that $\text{im}\,(C'(\lambda_o)) = F$. The statement follows now from (3).

Recall the examples for a-regular operators in section (2), in particular 3.(1). The respective even more specialized corollary resembles to the situation in [11], however, dispensing with the compactness of B.

(8) Let A', A'_h be a-regular, (B'_h) discretely compact with $B'_h \to B'(h \in H)$. If λ_0 is not an eigenvalue of (A',B'), then λ_0 is not a bifurcation point of (A,B).

Our aim is to show that at least certain eigenvalues of (A',B') are bifurcation points. A main tool will be provided by the topological degree for linear mappings. However, in infinite dimensional spaces the degree exists only for compact perturbations of the identity. Using the a-regularity and the Brouwer degree for the approximating operators we succeed in evading these difficulties by the following auxiliary result (cf. [11]). deg denotes the Brouwer degree and $\deg_{LS}$ the Leray-Schauder degree [1].

(9) Suppose the following conditions hold for $L-K \in B(E,F)$ and $L_h-K_h \in B(E_h,F_h)$:

(i) L, L_h and $L-K$, L_h-K_h a-regular

(ii) L and $L-K$ injective, K compact.

Then there exists a final piece $H_1 \subset H$ such that there holds

$$\deg(L_h-K_h, K_h(0,r), 0)$$
$$= \deg(L_h, K_h(0,r), 0)\deg_{LS}(I-KL^{-1}, L(K(0,r)), 0), h \in H_1.$$

Proof: From L, L_h a-regular and L injective there follows with 3.(8) that L is a homeomorphism and hence $KL^{-1} \in B(F)$ exists and is compact. Since we can write with $y=Lx$

$$0 = y - KL^{-1}y = Lx - Kx \Rightarrow x = 0,$$

from (ii) we get $I-KL^{-1}$ injective. In a similar manner we conclude from 3.(9) the existence of $L_h^{-1} \in B(F_h, E_h)$ for all indices h in a final piece $H_1 \subset H$. Hence, all operators in (iii) are well defined. Let us claim that $I-KL^{-1}$, $I_h-K_hL_h^{-1}$ is a-regular. Namely if $y_h \in F_h$ is a bounded sequence for any $H \subset H_0$ such that $(I_h-K_hL_h^{-1})y_h \to v \in F(h \in H)$ we can write as L and $L_h, h \in H_1$, are homeomorphisms, with the bounded sequence $x_h \equiv L_h^{-1}y_h \in E_h$, that $L_hx_h-K_hx_h \to v(h \in H_1')$. Hence, we conclude form (i) that there are $H_1'' \subset H_1'$ and $x \in E$ such that $x_h \to x(h \in H_1'')$ and $Lx-Kx=v$. Defining $y \equiv Lx$ this implies $y_h \to y(h \in H_1'')$ and $y-KL^{-1}y = v$. Using the injectivity of $I-KL^{-1}$, we can now deduce from 3.(8) that $I-KL^{-1}$ and $I_h-K_hL_h^{-1}$ are homeomorphisms for all $h \in H_2 \subset H_1$, H_2 a final piece. Moreover, 3.(8) delivers the convergence

$$(I_h-K_hL_h^{-1})^{-1} \to (I-KL^{-1})^{-1} (h \in H_2).$$

From these observations there follows

$$(10) \quad \deg(I_h-K_hL_h^{-1}, L_h(K_h(0,r)), 0) = \deg_{LS}(I-KL^{-1}, L(K(0,r)), 0).$$

In order to create a relation between the degrees of $I_h-K_hL_h^{-1}$ and L_h-K_h we define the homotopies

(11) $H_h(x_h,t) \equiv t(L_hx_h-K_hx_h)+(1-t)(I_h-K_hL_h^{-1})L_hx_h, \ h \in H_2.$

Assume that there exists a subsequence $H_2' \subset H_2$ such that

(12) $x_h \in S_h(0,r), t_h \in [0,1] : H_h(x_h,t_h) \to 0 (h \in H_2').$

For sake of simplicity we may assume without loss of generality that $t_h \to t \in [0,1] (h \in H_2')$ and introduce that value into (12),

$$t(L_hx_h-K_hx_h)+(1-t)(I_h-K_hL_h^{-1})L_hx_h \to 0 (h \in H_2').$$

That means

$$L_hx_h - K_hx_h \to 0 (H \in H_2').$$

Using (i), one has the existence of $H_2'' \subset H_2'$ and $x \in E$ such that

$$x_h \to x (h \in H_2''), \ Lx - Kx = 0.$$

From (2), we see $x \in S(0,r)$. However, this is a contradiction to L-K injective.

Hence, $H_h(x_h,t) \neq 0$ for all $x_h \in S_h(0,r), t \in [0,1]$. Introducing the homotopy theorem for the Brouwer degree, we get

(13) $\deg(L_h-K_h,K_h(0,r),0) = \deg((I_h-K_hL_h^{-1})L_h,K_h(0,r),0), h\in H_2.$

The last step consists in an application of the product formula for the Brouwer degree,

$$\begin{aligned} &\deg((I_h-K_hL_h^{-1})L_h, K_h(0,r),0) \\ (14) \quad &= \deg(L_h,K_h(0,r),0)\deg(I_h-K_hL_h^{-1}, L_h(K_h(0,r)),0), h \in H_2. \end{aligned}$$

Combining the relations (10), (13) and (14) we finally get the desired result.

Comparing the assupmtions of (9) and 3.(1), 3.(7) the following corollary is immediate.

(15) Suppose the following conditions hold for $L-K \in B(E,F)$, $L_h-K_h \in B(E_h,F_h)$:

(i) L,L_h a-regular

(ii) L and $L-K$ injective

(iii) F separable, $K_h \to K$ convergent and (K_h) discretely compact.

Then the assumptions of (9) are satisfied.

Proof: As in the proof of (9) we get the existence of $KL^{-1} \in B(F)$ and the existence of $K_hL_h^{-1} \in B(F_h)$ for all $h \in H_1 \subset H$. From (i) and (iii) there follows by 3.(1) immediately that $L-K, L_h-K_h$ is a-regular. The convergence $K_h \to K$ implies [12] the consistency of K,K_h, so that using 3.(7) there follows K compact.

A sequence $T_h:D(T_h)\ \ E_h \to F_h$ is called *equidifferentiable* for a sequence $u_h \in D(T_h)$, if T_h has the Fréchet derivative

$T'_h(u_h)$ at $u_h, h \in H_0$, and for each $\varepsilon>0$, there exists a $\delta>0$ such that

$$\| T_h(u_h+z_h) - T_h u_h - T'_h(u_h) z_h \| \leq \varepsilon \| u_h \|$$

for all z_h with $u_h+z_h \in D(T_h)$ and $\| z_h \| \leq \delta$, $h \in H_0$.

We can announce now our main result.

(16) Suppose that the following conditions hold:
(i) (A,A_h), $(C(\lambda), C_h(\lambda))$, and $(C'(\lambda), C'_h(\lambda))$ a-regular for all λ in an open set $G \subset \mathbb{R}$
(ii) $(C_h(\lambda))$ equidifferentiable at 0 and (B_h) maps bounded sequences into bounded sequences.
(iii) A' injective and B' compact.
If λ_0 is an eigenvalue of (A',B') with odd algebraic multiplicity, then λ_0 is a bifurcation point of (A,B).

Proof: Using (1) and 3.(8) we state that A' is a homeomorphism between E and F. Since K' compact we know that the spectrum of $C'(.)$ consists of a discrete set of eigenvalues. Hence we can find $\tau>0$ such that $[\lambda_0-\tau, \lambda_0+\tau]$ is in G and does not contain any other eigenvalue of (A',B') than λ_0. Choose $\beta \in (0,\tau)$. Since $\lambda_0+\beta$ is not an eigenvalue of (A',B') the operator $C'(\lambda_0+\beta)$ is injective. Observing (i) we conclude from 3.(9) that there exists a $\gamma>0$ and a final piece $H_1 \subset H$ such that

(17) $\gamma \| x \| \leq \| C'_h(\lambda_0+\beta) x_h \| \quad \forall x \in E$

and

(18) $\gamma \| x_h \| \le \| C'_h(\lambda_0+\beta)x_h \| \quad \forall x_h \in E_h, \; h \in H_1.$

Utilizing (ii), one has $\delta>0$ such that for all $x_h \in B_h(0,\delta)$ there holds

(19) $\| C_h(\lambda_0+\beta)x_h - C'_h(\lambda_0+\beta)x_h \| \le \frac{1}{2}\gamma \| x_h \|, \; h \in H_1,$

moreover, from 2.(3)

(20) $\| C(\lambda_0+\beta)x - C'(\lambda_0+\beta)x \| \le \frac{1}{2}\gamma \| x \|$

for all $x \in B(0,\delta)$. In order to get a relation between the mappings C_h and C'_h permitting the application of degree arguments we consider the homotopies

$$H_h(x_h,t) \equiv C'_h(\lambda_0+\beta)x_h + t(C_h(\lambda_0+\beta)x_h - C'_h(\lambda_0+\beta)x_h), \; h \in H.$$

Introducing (18) and (19) into an obvious estimate there follows for any $x_h \in S_h(0,\rho)$ with arbitrarily chosen $\rho \in (0,\delta)$

$$\| H_h(x_h,t) \| \ge \| C'_h(\lambda_0+\beta)x_h \| - \| C_h(\lambda_0+\beta)x_h - C'_h(\lambda_0+\beta)x_h \| \ge \frac{1}{2}\gamma\rho, h \in H_1.$$

This means that $H_h(x_h,t) \neq 0$ for all $x_h \in S_h(0,\rho)$ and $t \in [0,1]$ for all $h \in H_1$. The homotopy theorem for the Brouwer degree delivers now

$$\deg(C_h(\lambda_0+\beta), K_h(0,\rho), 0) = \deg({}'_h(\lambda_0+\beta), K_h(0,\rho), 0), h \in H_1.$$

A review of the assumptions and the preceding observations shows that Lemma (9) can be applied to $C'(\lambda_0+\beta) = (\lambda_0+\beta)B'-A'$ with the result

$$\deg(C_h(\lambda_0+\beta),K_h(0,\rho),0)$$
$$= \deg(A_h',K_h(0,\rho),0)\ \deg_{LS}(I-(\lambda_0+\beta)B'A'^{-1},A'(K(0,\rho)),0),$$

where $h \in H_2$, $H_2 \subset H$ a final piece. Using a theorem of Krasnoselski [9] that states

$$\deg_{LS}(I-(\lambda_0+\beta)B'A'^{-1},A'(K(0,\beta)),0) = (-1)^a,$$

we have now approached at

$$(21)\quad \deg(C_h(\lambda_0+\beta),K_h(0,\rho),0) = (-1)^a \deg(A_h',K_h(0,\rho),0),\quad h \in H_2.$$

Going again through the preceding part of the proof it becomes obvious that there can simultaneously be shown

$$(22)\quad \deg(C_h(\lambda_0-\beta),K_h(0,\rho),0) = (-1)^b \deg(A_h'(0,\rho),0),\quad h \in H_2.$$

Referring again to Krasnoselski [9], we introduce now that the difference a-b is equal to the algebraic multiplicity of λ_0, i.e., a differs form b with an odd number. Thus, by (21) and (22)

$$\deg(C_h(\lambda_0+\beta),K_h(0,\rho),0) \neq \deg(C_h(\lambda_0-\beta),K_h(0,\rho),0),\quad h \in H_2.$$

Using the homotopy property of the Brouwer degree inversely

we get that there are $t_h \in [0,1]$ and $x_h \in S_h(0,\rho)$ such that

$$(1-t_h)C_h(\lambda_0+\beta)x_h + t_h C_h(\lambda_0-\beta)x_h = 0, \quad h \in H_2.$$

That means

$$C_h(\lambda_0+\beta-2t_h\beta)x_h = 0, \quad h \in H_2.$$

Again we may assume without loss of generality that $t_h \to t \in [0,1] (h \in H_2)$ and insert this parameter value,

$$C_h(\lambda_0+\beta-2t\beta)x_h = 2\beta(t_h-t)B_h x_h, \quad h \in H_2.$$

The right side converges with $h \in H_2$ to 0 such that, using the a-regularity of $C(\lambda_0+\beta-2t\beta)$, $C_h(\lambda_0+\beta-2t\beta)$, we have a subsequence H_2^1 and $x \in E$ with

$$(23) \quad x_h \to x (h \in H_2^1), \quad (C(\lambda_0+\beta)-2t\beta)x = 0.$$

Since $x_h \in S_h(0,\rho)$, from (2) there follows $\|x\|=\rho$. Finally, regarding (17) and (20) one gets

$$\|C(\lambda_0+\beta)x\| \geq \|C'(\lambda_0+\beta)x\| - \|C'(\lambda_0+\beta)x-C(\lambda_0+\beta)x\| \geq \frac{1}{2}\gamma\|x\|,$$

what means, that $C(\lambda_0+\beta)x = 0$ holds only for $x=0$. The same consideration is valid for $C(\lambda_0-\beta)$. In view of (23) being true for $\|x\| = \rho$ there follows $t \in (0,1)$. As we have shown that for sufficiently small but arbitrarily chosen β there exists a $\delta>0$ such that, for every $\rho \in (0,\delta)$, the eigenvalue problem 2.(1) has a solution $(\lambda, x(\lambda))$ with $\lambda_0 \equiv \lambda + \beta - 2t\beta$ and $\|x(\lambda)\| = \rho$.

Let us add the usual specializing corollary.

(24) Suppose that the following conditions hold:
(i) $C(\lambda), C_h(\lambda)$ and A', A_h' a-regular
(ii) $(C_h(\lambda))$ equidifferentialble at ρ, (B_h') discretely compact and $B_h \to B$ convergent
(iii) F separable and A' injective.
If λ_0 is an eigenvalue of (A',B') with odd algebraic multiplicity, then λ_0 is a bifurcation point of (A,B).

Proof: Using 3.(1) and 3.(7) the hypothesisses of (16) are easily verified.

Remark, that B and B' compact must not be required explicitly.

5. APPROXIMATION OF BIFURCATION POINTS

The main advantage of our existence theory lies in its constructive approach permitting to introduce the general convergence theory for the approximate calculation of those eigenvalues of the linearized problems (A',B') which were identified previously as bifurcation points of (A,B).

Beginning with a corollary to 3.(8) we review some known convergence results for linear eigenvalue problems 2.(4) and their approximations 2.(8). For this purpose we have to interprete our real conditions as complex ones, in particular we will admit complex values of the parameter. In the following $\rho(A',B')$ and $\sigma(A',B')$ denote the *resolvent set* and the *spectrum* of (A',B') respectively. For all details and proofs we refer to [4,6].

(1) Let $C'(\lambda), C_h'(\lambda)$ be a-regular for each $\lambda \in \rho(A',B')$ and assume that $B_h' \to B'(h \in H_1)$. If $K \subset \rho(A',B')$ is compact, then there exists a final piece $H_1 \subset H$ such that $K \subset \rho(A_h',B_h')$.

This result implies that the limits λ of a subsequence of eigenvalues $\lambda_h \in \sigma(A_h',B_h')$ belong to the spectrum $\sigma(A',B')$ of $C'(.)$.

Let σ be a compact spectral set of (A',B'). If Δ is a Cauchy domain with boundary Γ, then

$$E(\sigma) \equiv \frac{1}{2\pi i} \int_{\Gamma} C'(\lambda)^{-1}B'd\lambda$$

is called the *spectral projection* of (A',B') with respect to σ. If σ consists of a finite number of poles of $C'(.)^{-1}B'$ then the range of $E(\sigma)$ equals the direct sum of the linear hulls of the respective sets of principal vectors.

What concerns the computation of $\sigma(A',B')$ using the approximating spectra $\sigma(A_h',B_h')$ the following result has been shown.

(2) <u>For an isolated point</u> λ_o <u>of</u> $\sigma(A',B')$ <u>and a related Cauchy domain</u> Δ <u>with boundary</u> Γ <u>holds</u>

(3) $\inf\{\text{dist}(\mu,\lambda_o)\,|\,\mu\in\sigma(A_h',B_h')\}\to 0(h\in H_o)$.
<u>For a final piece</u> $H_1\subset H$ <u>is</u> $\Gamma\subset\rho(A_h',B_h')$, <u>so there exist</u> $E_h(\sigma_h)$, <u>where</u> $\sigma_h\equiv\sigma(A_h',B_h')$, $h\in H_1$.
<u>Finally there holds</u>

$$\liminf \text{rank}\, E_h(\sigma_h)\geq \text{rank}\, D(\lambda_o).$$

<u>For some final piece there is</u> $\sigma \neq \emptyset$.

In view of our preceding considerations the following theorem enjoys particular importance.

(4) <u>For an isolated point</u> λ_o <u>of</u> $\sigma(A',B')$ <u>the following conditions are equivalent</u>:
(i) $\text{rank}\, E(\lambda_o)<\infty$, $\text{rank}\, E_h(\sigma_h)\to\text{rank}\, E(\lambda_o)(h\in H_1)$
(ii) $\text{im}(E(\lambda_o))$ <u>is separable and the sequence</u> $(E_h(\sigma_h))$ <u>is discretely compact</u>.
<u>If one of them holds then</u> λ_o <u>is a pole of</u> $C(.)^{-1}$.
<u>For a final piece</u> H_2 <u>the set</u> σ_h <u>consists of</u> $m(\sigma_h)=\text{rank}\, E(\lambda_o)$ <u>algebraically repeated eigenvalues</u> $\mu_h^{(1)},\ldots,\mu_h^{(m)}$ <u>with</u>

$$\mu_h^{(j)}\to\lambda_o(h\in H_2),\ j=1,\ldots,m.$$

The application of these theorems to the problem discussed previously is nearly obvious. We have to take the sequence of approximating linearized problems 2.(8).

Under the assumptions of (4) which are satisfied by the hypotheses of our specialized corollaries there have to be determined those approximating spectral sets σ_h of (A_h', B_h') with an odd joint algebraic multiplicity $m(\sigma_h)$ for all $h \in H_2$. In view of (1), (3), the existence of σ_h is clear. Their limits are the desired bifurcation points of (A,B).

REFERENCES

[1] Deimling, K., Nichtlineare Gleichungen und Abbildungsgrade, Hochschultext, Springer 1974.

[2] Grigorieff, R. D., Zur Theorie linearer approximationsregulärer Operatoren, I und II, Math. Nachr. 55, 233-249, 251-263 (1973)

[3] Grigorieff, R. D., Über diskrete Approximationen nichtlinearer Gleichungen 1, Art. Math. Nachr. 69, 253-272 (1975)

[4] Grigorieff, R. D., Diskrete Approximation von Eigenwertproblemen I, Numer. Math. 24, 355-374 (1975)

[5] Grigorieff, R. D., Jeggle, H., Approximation von Eigenwertproblemen bei nichtlinearer Parameterabhängigkeit, manuskripta math. 10, 245-271 (1973)

[6] Jeggle, H., Über die Approximation von linearen Gleichungen zweiter Art und Eigenwertproblemen in Banachräumen, Math. Z. 124, 319-342 (1972)

[7] Jeggle, H., Zur Störungstheorie nichtlinearer Variationsungleichungen, Lecture Notes in Math. 477, 158-176, Springer 1975

[8] Jeggle, H., Wendland, W., On the discrete approximation of eigenvalue problems with holomorphic parameter dependence, Proc. Royal Soc. Edinburgh 78A, 1-29 (1978)

[9] Krasnoselski, M. A., Topological methods in the theory of nonlinear integral equations, Pergamon Press 1964

[10] Petryshyn, W. V., The approximation-solvability of equations involving A-proper and pseudo-A-proper mappings, Bull, AMS 81, 223-312 (1975)

[11] Petryshyn, W. V., Bifurcation and asymptotic bifurcation for equations involving a-proper mappings with applications to differential operators, Preprint 1976

[12] Stummel, F., Diskrete Konvergenz linearer Operatoren I., Math. Ann. 190, 45-92 (1970)

[13] Wolf, R., Über lineare approximationsreguläre Operatoren, Math. Nachr. 59, 325-341 (1974)

CALCULATING GLOBAL BIFURCATION

M. Prüfer

Institut für Angewandte Mathematik
Universität Bonn
Bonn, Federal Republic of Germany

I. INTRODUCTION

The main purpose of this paper is to introduce an idea how to make local and global bifurcation in the sense of M.A. Krasnosel'skii [9] and P.H. Rabinowitz [12] constructive. For our approach it is important to remark that the main tools tools used in the arguments of the papers of Krasnosel'skii and Rabinowitz are: the topological fixed point index and the continuation method of Poincaré. In view of this, an appropriate numerical translation of both is relevant. This aim is achieved by using Sperner's Lemma essentially.

[1] *This work was supported by the Deutsche Forschungsgemeinschaft, Sonderforschungsbereich 72, Universität Bonn.*

ISBN 0-12-729250-0

In section II we list the main tools used in this paper. In section III we outline the main ideas concerning the approximation of fixed points by Sperner-simplices. In view of our attempt mentioned above, the main result of this section is a simplicial interpretation of the fixed point index by means of Sperner-simplices. In section IV we describe a simplicial version of the Poincaré continuation method. Combining the ideas of sections III and IV, section V is our main section and is devoted to our numerical approach to global bifurcation. In section VI we discuss our numerical experience concerning the approximation of "unstable" periodic solutions of an autonomous functional differential equation of type

$$\dot{x}(t) = -\lambda f(x(t-1)).$$

The results presented in this paper are part of the author's dissertation to be submitted to the University of Bonn.

I would like to thank Professor Heinz-Otto Peitgen for his advice and many helpful discussions. He inspired my interest in the subject and contributed much to the topological insight.

II. PRELIMINARIES AND NOTATIONS

A set of vectors $\{v^0,\dots,v^k\}$ lying in some $\mathbb{R}^n$ will be called independent, if the set of vectors $\{(v^0-v^1),\dots,(v^0-v^k)\}$ is linearly independent. The closed convex hull of a set of independent vectors will be called a k-simplex and will be denoted by $\langle v^0,\dots,v^k\rangle$. The number k will be called the dimension of the simplex. Let e^0 denote the origin in $\mathbb{R}^n$, and let $\{e^1,\dots,e^n\}$ be the standard basis in $\mathbb{R}^n$. The n-simplex $S := \langle e^0,\dots,e^n\rangle$ will be called the unit-n-simplex.

Let $\langle v^0,\dots,v^k\rangle$ be a k-simplex, and let $\{v^{i_0},\dots,v^{i_p}\} \subset \{v^0,\dots,v^k\}$ be a subset, where the v^{i_j} are assumed to be mutually different. The p-simplex $\langle v^{i_0},\dots,v^{i_p}\rangle$ will be called a face of $\langle v^0,\dots,v^k\rangle$.

Let $\langle v^0,\dots,v^k\rangle$ be a k-simplex, and let $\{\sigma_1^k,\dots,\sigma_t^k\}$ be a finite collection of k-simplices satisfying

(2.1) $\bigcup_{1\le j\le t} \sigma_j^k = \langle v^0,\dots,v^k\rangle$, and,

(2.2) if $\sigma_i \cap \sigma_j \neq \emptyset$, then $\sigma_i \cap \sigma_j$ is a common face of σ_i and σ_j.

A collection of simplices like the above together with all its faces will be called a triangulation of $\langle v^0,\dots,v^k\rangle$ and abbreviated by τ. A p-simplex which belongs to a triangulation will always be denoted by σ^p. A 0-simplex will be called a <u>vertex</u>.

If $\langle v^o,\dots,v^n\rangle$ is an n-simplex, a function ℓ: $\langle v^o,\dots,v^n\rangle \to N_o := \{0,\dots,n\}$ will be called a <u>proper labeling</u> if

(2.3) $\{\ell(v^o),\dots,\ell(v^n)\} = N_o$, and,
(2.4) if $x \in \langle v^{i_o},\dots,v^{i_k}\rangle$, then $\ell(x) \in \{\ell(v^{i_o}),\dots,\ell(v^{i_k})\}$.

Let τ be a triangulation of a properly labeled n-simplex.

A simplex $\sigma^n = \langle a^o,\dots,a^n\rangle \in \tau$ will be called <u>completely labeled</u> or a <u>Sperner-simplex</u>, if $\{\ell(a^o),\dots,\ell(a^n)\} = N_o$. The collection of all Sperner-simplices in a triangulation τ will be denoted by $N(\tau,\ell)$. The following result, which was first proved in [14], will be basic for our considerations:

<u>Sperner's Lemma:</u> Let $\langle v^o,\dots,v^n\rangle$ be an n-simplex and let ℓ: $\langle v^o,\dots,v^n\rangle \to N_o$ be a proper labeling. Then $\# N(\tau,\ell)$ is an odd number for any triangulation τ of $\langle v^o,\dots,v^n\rangle$.

Next we introduce orientations of Sperner-simplices. Let $\langle v^o,\dots,v^n\rangle \subset \mathbb{R}^n$ be a properly labeled n-simplex. Let τ be a triangulation of $\langle v^o,\dots,v^n\rangle$ and let $\sigma^n = \langle a^o,\dots,a^n\rangle \in \tau$ be a Sperner-simplex. Assume without loss of generality that $\ell(v^i) = \ell(a^i) = i$ for $0\le i\le n$. There is a unique affine homeomorphism $\mu = M + m : \mathbb{R}^n \to \mathbb{R}^n$ (M an $(n\times n)$-matrix and $m \in \mathbb{R}^n$), such that $\mu(a^i) = v^i$ for all $0\le i\le n$. If $\det M \underset{(<)}{>} 0$, σ^n will be called _positively_ (_negatively_) _oriented_ (notation: $or(\sigma^n) = \overset{+}{(-)}1$). We have the following

Oriented version of Sperner's Lemma: Let $\langle v^o,\ldots v^n\rangle$ be an n-simplex and let $\ell: \{v^o,\ldots,v^n\} \to N_0$ be a proper labeling. Then we have

$$\sum_{\sigma^n \in N(\tau,\ell)} or(\sigma^n) = 1$$

for any triangulation τ of $\langle v^o,\ldots,v^n\rangle$.

Let $P \subset \mathbb{R}^n$ be a polyhedron, and let $U \subset P$ be an open set. If $f{:}U \to P$ is a continuous mapping, then Fix(f) denotes the set of all fixed points of f. If $\partial U \cap \mathrm{Fix}(f) = \emptyset$, and if Fix(f) is a compact set, then the _(topological) fixed point index_ i(f,U,P) is defined (for the definition and basic properties cf. [3]).

If C is a component of fixed points of f, the integer i(f,U,P) will be the same for all open neighbourhoods $U \subset P$ of C with $\overline{U} \cap \mathrm{Fix}(f) = C$. This integer is called the _local fixed point index_ of f at C and will be denoted by i(f,C).

III. FIXED POINTS AND SPERNER-SIMPLICES

In this section we outline the main ideas concerning the approximation of fixed points by Sperner-simplices. These are intimately related to the attempts to give a constructive proof of the Brouwer Fixed Point Theorem. The first such proof was given in 1929 by B. Knaster, C. Kuratowski and S. Mazurkiewicz ([8]). However, in 1967 the first effective numerical procedures for searching Sperner-simplices were given by D. I. A. Cohen ([4]) and H. H. Scarf ([13]).

The basic idea, when using Sperner's Lemma to approximate fixed points, is the following:

Let $f:S \to S$ be a continuous mapping and let τ be a triangulation of S. Note that the fixed point set of f can look very wild. However, by a discretization process via τ, the mapping f can be replaced by a continuous mapping h which satisfies:

(3.1) The fixed point set of h consists of a finite number of isolated points and there is a one-to-one correspondence between Fix(h) and $N(\tau,\ell^f)$ (for the definition of ℓ^f cf. (3.4), (3.5)).

(3.2) Every fixed point of h is "close" to a component of fixed points of f.

(3.3) The fixed point sets of f and h are equivalent in a topological sense.

To describe the construction of h, we first introduce a proper labeling ℓ^f, which is induced by f:

For $x \in S$ and $1 \le i \le n$ we define:

$$(3.4) \quad L_i^f(x) = \begin{cases} 1 & \text{if } x_i = 0 & (3.5.1) \\ 1 & \text{if } f_i(x) > x_i & (3.5.2) \\ 0 & \text{if } f_i(x) \le x_i \ne 0 & (3.5.3) \end{cases}$$

We define:

$$(3.5) \quad \ell^f(x) = \text{initial number of 1's in the vector } (L_1^f(x),\ldots,L_n^f(x)).$$

One observes that $\ell^f : S \to N_0$ is a proper labeling. The next step consists in replacing Fix(f) by a somewhat more "regular" set $\mathcal{F} \subset \mathrm{Fix}(f)$:

Setting

$$(3.6) \quad A_i^f = \mathrm{cl}\{x \in S : \ell^f(x) = i\} \quad \text{for } 0 \le i \le n,$$

we have

$$(3.7.1) \quad \bigcup_{0 \le i \le n} A_i^f = S \quad \text{and} \quad (3.8.2) \quad \mathcal{F} := \bigcap_{0 \le i \le n} A_i^f \subset \mathrm{Fix}(f).$$

Easy examples show that the inclusion in (3.7.2) is proper. Fixed points in $\mathrm{Fix}(f)\setminus\mathcal{F}$ are "degenerate"; for example, they can disappear by a small perturbation of f. Thus, replacing the approximation problem Fix(f) by $\mathcal{F}$, means to consider a more "regular problem": for example, $\mathcal{F}$ contains no open set $\mathcal{O} \subset S$; however, $\mathcal{F}$ may contain lower dimensional continua, which need not even be manifolds; it will follow from theorem 1 below that any component $C \subset \mathrm{Fix}(f)$ with $i(f,C) \neq 0$ intersects F.

Now we are ready for the construction of h, which depends on the triangulation τ of S:

Define the following vectors in $\mathbb{R}^n$:

(3.8.1) $b^o = (1,0,...,0)$

and for $0<j<n$ define b^j by:

(3.8.2) $b_i^j = \begin{cases} -1 & \text{if } i = j \\ 1 & \text{if } i = j+1 \\ 0 & \text{otherwise.} \end{cases}$

Finally set:

(3.8.3) $b^n = (0,...,0,-1)$.

Let y be a vertex of τ with $\ell^f(y) = i$. Set $h(y) = y - \varepsilon b^i$, $\varepsilon > 0$. Then define h as the barycentric extension on each simplex σ^n in τ. To obtain that $h(S) \subset S$, one chooses ε sufficiently small.

From the definition of h we have:

(3.9) $x \in \mathrm{Fix}(h) \Rightarrow$ $x \in \mathrm{int}\, \sigma^n$, where σ^n is an ℓ^f-Sperner-simplex.

Moreover, each $\sigma^n \in \tau$ contains at most one fixed point of h; i.e. h has property (3.1).

Furthermore, h satisfies (3.2): for any neighbourhood U of F in S there exists a $\delta > 0$ with the following property:

if τ is any triangulation of S with mesh$\tau \leq \delta$, and if $\sigma^n \in \tau$ is an ℓ^f-Sperner-simplex, then $\sigma^n \subset U$.

To comment on property (3.3) we note that if x is a fixed point of h, we have $i(h,x) = i(h,\text{int}\sigma^n,S) = \text{or}(\sigma^n)$, where $\sigma^n \in \tau$ is the ℓ^f-Sperner-simplex with $x \in \sigma^n$ (cf. 3.9). Thus, by the additivity property of the index, we have for any set U which is the union of n-simplices $\sigma^n \in \tau$:

$$(3.1o)\quad i(h,\text{int } U,S) = \sum_{\substack{\sigma^n \in N(\tau,\ell^f) \\ \sigma^n \subset U}} \text{or}(\sigma^n).$$

In particular, if U = S in (3.1o) we obtain by the normalization property of the index:

$$(3.11)\quad \sum_{\sigma^n \in N(\tau,\ell^f)} \text{or}(\sigma^n) = 1.$$

Thus, (3.1o) may be regarded as a "localization" of the oriented version of Sperner's Lemma and (3.11) appears in our context as a special case of the Lefschetz-Hopf-Theorem (cf. [3]). The following theorem, which is proved in [11], asserts that such a "localization" is also possible for our original mapping f and this is what we have to make precise in (3.3).

Theorem 1: Let S be the unit-n-simplex, and let $f:S \to S$ be a continuous mapping. Let τ be a triangulation of S, and let U be a union of n-simplices belonging to τ. If mesh τ is sufficiently small, we have:

(3.12) $i(f, \text{int } U, S) = i(h, \text{int } U, S)$,

where h is the mapping considered above.

Together with (3.1o), equality (3.12) is fundamental for our further considerations and provides the essential tool to attack bifurcation problems numerically by means of simplicial topology.

Corollary 1 answers the question: which elements of Fix(f) can be approximated by Sperner-simplices:

<u>Corollary 1:</u> Let $f: S \to S$ be a continuous mapping, and let $\{\tau_k\}_{k \in \mathbb{N}}$ be some sequence of triangulations of S with $\lim_{k\to\infty} \text{mesh}\,\tau_k = 0$. Let C be a component of Fix(f), and let $U \subset S$ be an open neighbourhood of C, such that $\overline{U} \cap \text{Fix}(f) = C$. Then there exists a $k_o \in \mathbb{N}$ such that for all $k \geq k_o$ τ_k contains at least $|i(f,C)|$ Sperner-simplices in U.

Corollary 1 is similar to a result of B. A. Vertgeim [16]. In contrast to [16], our result gives a precise lower bound for the number of Sperner-simplices, which approximate a component of fixed points.

For our further considerations it is important to get rid of the restriction that f maps S into itself. We extend the notion of a labeling function:

If $f: U \to \mathbb{R}^n$ is continuous, where $U \subset \mathbb{R}^n$ is a union of n-simplices, we obtain an appropriate labeling $\ell^f: U \to N_0$ just be omitting condition (3.4.1) in (3.4). Assuming that Fix(f) is compact and that $\partial U \cap \mathrm{Fix}(f) = \emptyset$ we can extend (3.12):

$$(3.13)\quad i(f, \mathrm{int}\, U, \mathbb{R}^n) = \sum_{\sigma^n \in N(\tau, \ell^f)} \mathrm{or}(\sigma^n),$$

where τ is any triangulation of U, which is sufficiently fine.

Aside from corollary 1, theorem 1 gives an intuitive idea of

a) How to construct for a prescribed integer k a mapping f having an isolated fixed point x^* such that $i(f, x^*) = k$.

b) How to find (in a neighbourhood of x^*) a perturbation $\tilde{f}$ of f and fixed points $x_1^*, \ldots, x_p^*$ of $\tilde{f}$ such that $|i(\tilde{f}, x_j^*)| = 1$ for all $1 \le j \le p$ and $\sum_{j=1}^{p} i(\tilde{f}, x_j^*) = k$.

This is demonstrated by the following examples:

(3.14.1)

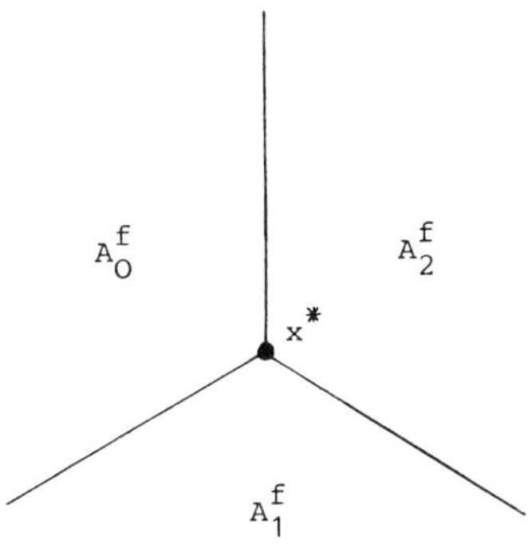

(3.14.2)

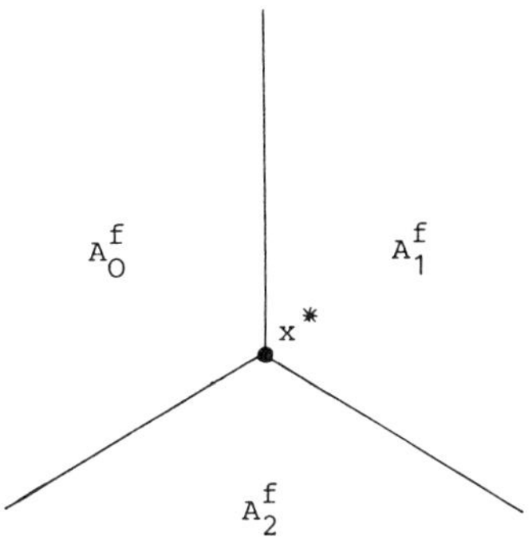

Any fixed point of type (3.14.1) (resp. (3.14.2)) has local index +1 (resp. -1).

(3.15.1)

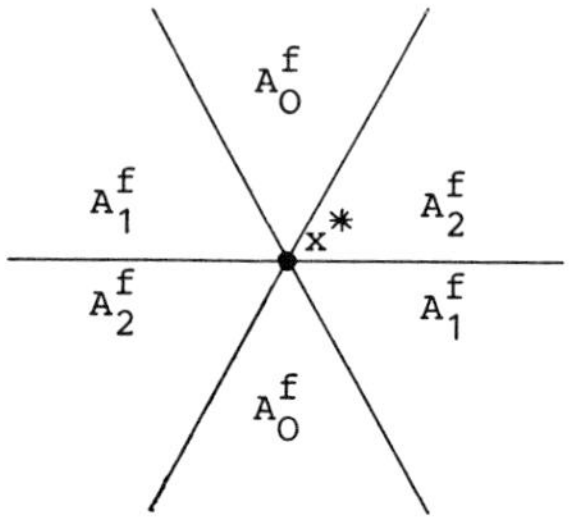

(3.15.2)

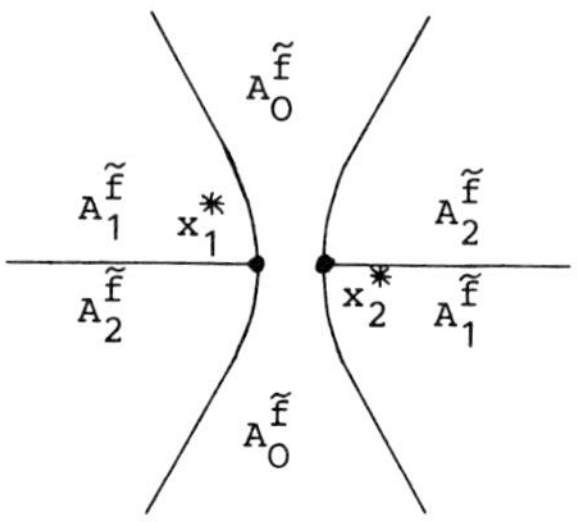

The fixed point x^* in (3.15.1) has local index +2. By a perturbation $\tilde{f}$ of f, two fixed points x_1^*, x_2^* may be produced, each of which is of type (3.14.1). Thus
$i(\tilde{f},x_1^*) = i(\tilde{f},x_2^*) = 1.$

(3.16.1)

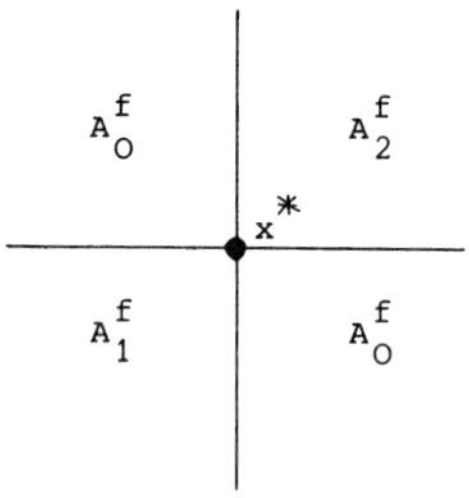

(3.16.2)

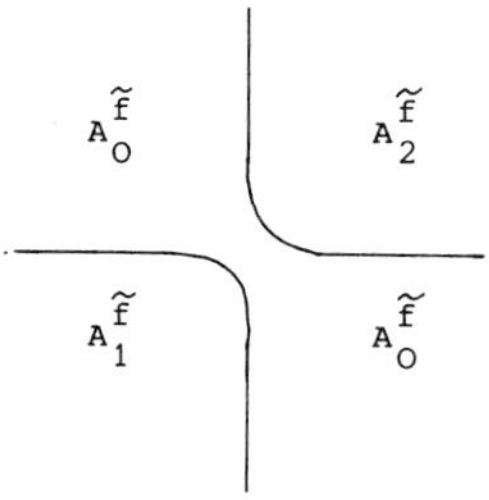

(3.16.3)

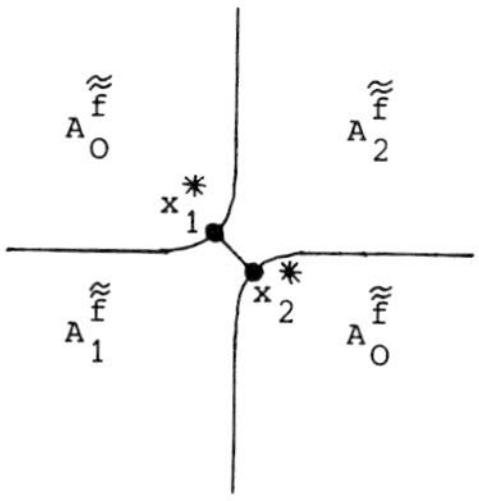

The fixed point x^* in (3.16.1) has local index zero. Figure (3.16.2) shows how a fixed point can disappear. Also, a different perturbation $\tilde{\tilde{f}}$ can generate two fixed points x_1^* and x_2^* (cf. 3.16.3). The fixed point x_1^* is of type

(3.14.1), x_2^* is of type (3.14.2). Thus,

$1 = i(\tilde{\tilde{f}}, x_1^*) = -i(\tilde{\tilde{f}}, x_2^*)$.

The observations (3.14) - (3.16) seem to be of some interest in the context of questions arising in Nielsen-Wecken theory (cf. [3]).

IV. THE POINCARÉ CONTINUATION METHOD

Since Scarf's work [13], procedures of various types for searching Sperner-simplices have been developed, such as homotopy techniques, restart techniques, and many others (for a survey cf. [1], [15]).

In this section we will describe a procedure, which is similar to a homotopy algorithm given by B. C. Eaves and R. Saigal ([5], [6]).

This procedure can be regarded as a simplicial version of what is often called the "Poincaré continuation method" (cf. [7]):

A fixed point problem $f(x) = x$ with a continuous mapping $f: S \to S$ in general is very hard to solve. Instead of the original problem consider a problem $g(x) = x$, which is easy to solve (for example, choose g to be a constant mapping). Since S is convex, f and g are homotopic. Choose a continuous homotopy $H: S \times [0,1] \to S$, which joins f and g. Then there exists a connected set $C \subset S \times [0,1]$ with $H(x,t) = x$ for all $(x,t) \in C$ and $C \cap S \times \{0\} \neq \emptyset \neq C \cap S \times \{1\}$.

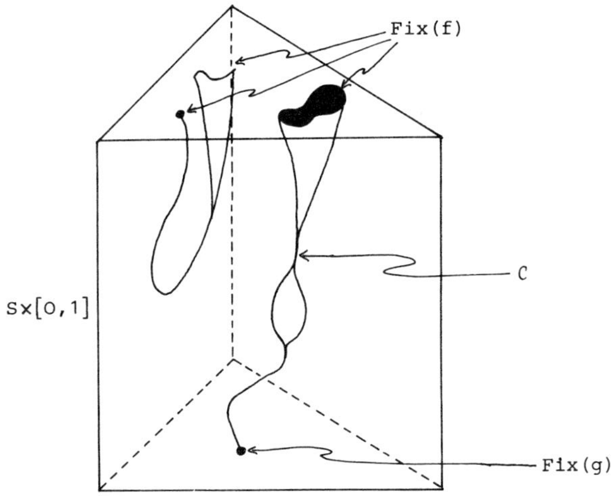

FIGURE 1.

The work that remains to be done is to follow the connected set C by some numerical method.

The first step in our approach is to define a labeling $\ell^H: S \times [0,1] \to N_0$: if $H: S \times [0,1] \to S$ is a continuous homotopy joining f and g, $H(.,t)$ is a continuous mapping from S into S for all $t \in [0,1]$. Thus, for all $t \in [0,1]$ we have by (3.4) and (3.5) a proper labeling $\ell^{H(.,t)}: S \to N_0$. We define

$$(4.1) \quad \ell^H(x,t) = \ell^{H(.,t)}(x)$$

for all $(x,t) \in S \times [0,1]$.

Let now τ be a triangulation of $S \times [0,1]$ and let σ_0^n be an ℓ^g-Sperner-simplex in $\tau|_{S \times \{0\}}$. By a process, which is described in [2] and [4], we can form a chain

$$(4.2) \quad \sigma_0^n, \sigma_0^{n+1}, \sigma_1^n, \sigma_1^{n+1}, \ldots, \sigma_p^n, \sigma_p^{n+1}, \sigma_{p+1}^n$$

of n- (resp. (n+1)-) simplices in τ with the following properties:

(4.3) Each n-simplex σ_i^n is completely labeled under the labeling ℓ^H,

(4.4) for $0<i\le p$ σ_i^n is a common face of σ_{i-1}^{n+1} and σ_i^{n+1},

(4.5) σ_0^n and σ_{p+1}^n are the only n-simplices in the chain, which are contained in $\partial(S \times [0,1])$. σ_0^n (resp. σ_{p+1}^n) is a face of σ_0^{n+1} (resp. σ_p^{n+1}).

A chain of type (4.2), which is started with an ℓ^g-Sperner-simplex $\sigma^n \in \tau|_{S \times \{0\}}$, will be denoted by $Ch(\sigma^n)$.

Since $\ell^{H(.,t)}$ is a proper labeling for each $t \in [0,1]$ and because of the oriented version of Sperner's Lemma, we have the following

Theorem 2: Let S be the unit-n-simplex and $f,g:S \to S$ continuous mappings. Let $H:S \times [0,1] \to S$ be an arbitrary continuous homotopy joining f and g. Let τ be a triangulation of $S \times [0,1]$. Then the number of ℓ^g-Sperner-simplices σ^n in $\tau|_{S \times \{0\}}$, such that $Ch(\sigma^n)$ ends with an ℓ^f-Sperner-simplex in $\tau|_{S \times \{1\}}$, is odd.

In particular, if $g:S \to S$ is chosen such that there exists a unique ℓ^g-Sperner-simplex in $\tau|_{S \times \{0\}}$, theorem 2 describes a procedure how to find an ℓ^f-Sperner-simplex in $\tau|_{S \times \{1\}}$ which approximates a fixed point of f.

With the same arguments as in section III, we can say that by introducing a triangulation in $S \times [0,1]$, the original homotopy is replaced by a piecewise linear homotopy. In particular, the possibly highly "irregular" set $\check{C} = \{(x,t) \in S \times [0,1] : H(x,t) = x\}$ (higher dimensional continua, bifurcations), is replaced by a 1-dimensional, piecewise linear manifold.

The following theorem asserts that this manifold approximates $\check{C}$ if the triangulation τ is sufficiently fine.

Theorem 3: Let S be the unit-n-simplex and $f,g:S \to S$ continuous mappings. Let $H:S \times [0,1] \to S$ be a continuous homotopy joining f and g. Let $\check{C} = \{(x,t) \in S \times [0,1] : H(x,t) = x\}$ and let $U \subset S$ be an open neighbourhood of $\check{C}$. Then there exists a $\delta > 0$, such that for all triangulations τ of $S \times [0,1]$ with mesh$\tau \le \delta$ we have

$$\bigcup_{\sigma^n \in N(\tau|_{S \times \{0\}}, \ell^g)} Ch(\sigma^n) \subset U.$$

V. CALCULATING GLOBAL BIFURCATION

From an abstract point of view there is no difference between a homotopy equation $H(x,t) = x$ and an eigenvalue problem

(5.1) $F(x,\lambda) = x$

where $F:\mathbb{R}^n \times J \to \mathbb{R}^n$ is a continuous mapping, and $J = [\lambda_1,\lambda_2]$, $\lambda_1 \neq \lambda_2$, is a compact interval of real numbers. The set of solutions of (5.1) will be denoted by S.

For each $\lambda \in [\lambda_1,\lambda_2]$ $F(.,\lambda)$ maps $\mathbb{R}^n$ into $\mathbb{R}^n$. Omitting condition (3.4.1), we obtain by (3.4) and (3.5) a labeling $\ell^{F(.,\lambda)}:\mathbb{R}^n \to N_0$. As in (4.1), we define for $(x,\lambda) \in \mathbb{R}^n \times J$:

(5.2) $\ell^{F}(x,\lambda) = \ell^{F(\cdot,\lambda)}(x)$

By a theorem which is analogous to theorem 2 we will describe a procedure for the approximation of a component of S.

We make the following assumption (cf. figure 2):

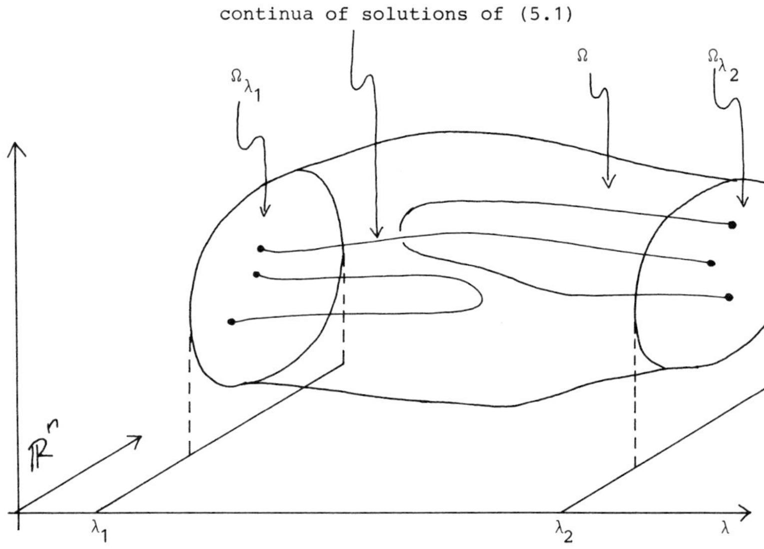

FIGURE 2.

(H) $\begin{cases} \text{There exists an open, connected set } \Omega \subset \mathbb{R}^n \times J \text{ such} \\ \text{that } \overline{\Omega} \text{ is compact, } S \cap \partial\Omega = \emptyset, \text{ and } \Omega_{\lambda_1} \neq \emptyset \neq \Omega_{\lambda_2}, \text{ where} \\ \Omega_\lambda = \Omega \cap \mathbb{R}^n \times \{\lambda\}. \end{cases}$

According to theorem 1 and theorem 2, we have the following

<u>Theorem 4:</u> Let $F:\mathbb{R}^n \times J \to \mathbb{R}^n$ be a continuous mapping. Assume that (H) holds and $i(F(.,\lambda_1),\Omega_{\lambda_1},\mathbb{R}^n \times \{\lambda_1\}) \neq 0$.

Then there exists a $\delta > 0$ such that each triangulation τ of $\mathbb{R}^n \times J$ with mesh$\tau \leq \delta$ has the following property:

There exist at least $|i(F(.,\lambda_1),\Omega_{\lambda_1},\mathbb{R}^n \times \{\lambda_1\})|$ ℓ^F-Sperner-simplices $\sigma_1^n,\ldots,\ \sigma_p^n$ in Ω_{λ_1} such that for each $i \in \{1,\ldots,p\}$ $Ch(\sigma_i^n)$ ends with an ℓ^F-Sperner-simplex in Ω_{λ_2}.

Theorem 4 guarantees the existence of a union of piecewise linear paths joining Ω_{λ_1} and Ω_{λ_2}. By a theorem which is analogous to theorem 3 we may assert that each of these paths "comes close" to the set S of solutions of (5.1), if the triangulation is sufficiently fine. Thus, in view of a numerical realization, theorem 4 gives an algorithm for the solution of an eigenvalue problem, provided ℓ^F-Sperner-simplices in Ω_{λ_1} can be detected.

In the situation discussed above, our assumption was $S \cap \partial\Omega = \emptyset$. Then the homotopy invariance of the index implies $i(F(.,\lambda_1),\Omega_{\lambda_1},\mathbb{R}^n \times \{\lambda_1\}) = i(F(.,\lambda_2),\Omega_{\lambda_2},\mathbb{R}^n \times \{\lambda_2\})$.

However, in topological bifurcation theory the situation $S \cap \partial\Omega \neq \emptyset$ is typical.

We consider the case $J = \mathbb{R}$, and we make the additional assumption that the operator F satisfies $F(0,\lambda) = 0$ for all $\lambda \in \mathbb{R}$.

A <u>bifurcation point</u> of (5.1) is a point $(0,\lambda^*)$, $\lambda^* \in \mathbb{R}$, such that each neighbourhood of $(0,\lambda^*)$ contains a point $(x,\lambda) \in S$ with $x \neq 0$. We quote the finite-dimensional case of the local bifurcation result of M. A. Krasnosel'skii [9]:

<u>Theorem 5:</u> Let $F:\mathbb{R}^n \times \mathbb{R} \to \mathbb{R}^n$ be a continuous operator, such that $F(0,\lambda) = 0$ for all $\lambda \in \mathbb{R}$. Suppose, we have $\lambda_1,\lambda_2 \in \mathbb{R}$, $\lambda_1 < \lambda_2$, such that zero is an isolated fixed point of $F(.,\lambda_i)$ $(i = 1,2)$. Suppose that $i(F(.,\lambda_1),0) \neq$ $\neq i(F(.,\lambda_2),0)$. Then there exists a bifurcation point $(0,\lambda^*)$ of (5.1) with $\lambda_1 \leq \lambda^* \leq \lambda_2$.

In view of theorem 5, we obtain the following simplicial interpretation:

<u>Theorem 6:</u> Let λ_1, λ_2 be as described in theorem 5. For $i = 1,2$ choose open neighbourhoods $\Omega_{\lambda_i} \subset \mathbb{R}^n \times \{\lambda_i\}$ of $(0,\lambda_i)$ such that zero is the only fixed point of $F(.,\lambda_i)$ in $\overline{\Omega}_{\lambda_i}$. Let τ be a triangulation of $\mathbb{R}^n \times \mathbb{R}$ such that $\{\sigma^p \in \tau : \sigma^p \subset \mathbb{R}^n \times \{\lambda_i\}\}$ is a triangulation of $\mathbb{R}^n \times \{\lambda_i\}$. Let mesh$\tau$ be sufficiently small.

Then there exists at least $|i(F(.,\lambda_1),0) - i(F(.,\lambda_2),0)|$ ℓ^F-Sperner-simplices $\sigma_1^n,\dots,\sigma_p^n$ in $\Omega_{\lambda_1} \cup \Omega_{\lambda_2}$ with the following property:

For each σ_j^n, $Ch(\sigma_j^n)$ bifurcates from the line of trivial solutions and follows a branch of nontrivial solutions.

Chains like those in theorem 6 obviously either tend to infinity in $\mathbb{R}^n \times \mathbb{R}$ or approach the parameter axis. This proves the finite-dimensional case of the global bifurcation result of P. H. Rabinowitz [12].

VI. NUMERICAL EXPERIENCE

The algorithm described in this paper has been successfully used in the approximation of "unstable" periodic solutions of an autonomous functional differential equation of type:

$$(6.1)\quad \begin{cases} \dot{x}(t) = -\lambda f(x(t-1))\,; & t \geq 0 \\ x(t) = \Phi(t)\,; & -1 \leq t \leq 0 \end{cases}$$

Equations like the above model population growth phenomena. In view of this, the following assumptions are natural:

(6.2) $f:\mathbb{R} \to \mathbb{R}$ is continuous

(6.3) $t \cdot f(t) > 0$ for all $t \neq 0$.

(6.4) f is differentiable in a neighbourhood U of zero. Assume $f'(0) = 1$ and $f'(t) > 1$ for all $t \in U \setminus \{0\}$.

We sketch the problem: For an initial value $\Phi:[-1,0] \to \mathbb{R}$ and a fixed value $\lambda \in \mathbb{R}_+$ there exists a unique continuous function $x(t;\Phi,\lambda):[-1,\infty) \to \mathbb{R}$ which is differentiable on $(0,\infty)$ and which is a solution of the initial value problem (6.1). The problem is now to find pairs (Φ,λ) such that $x(t;\Phi,\lambda)$ is periodic in t.

Theoretical studies of R. D. Nussbaum ([10]) suggest that for this problem one has to restrict the initial data Φ to the cone

$$P = \{\Phi:[-1,0] \to \mathbb{R} : \Phi \text{ continuous}, \Phi(-1) = 0, \Phi \text{ monotonicly increasing on } [-1,0]\}.$$

Using abstract fixed point theory, existence of solutions has been studied by considering the shift-operator

$$F:P \times \mathbb{R}_+ \to P$$

which is defined by

$$(6.5) \quad F(\Phi,\lambda) = \begin{cases} x(z_2+1+t;\Phi,\lambda) & \text{if } z_2, \text{ the second zero of } x(t;\Phi,\lambda) \text{ exists} \\ 0 & \text{otherwise.} \end{cases}$$

According to the definition of the nonlinear operator F, we have that solutions of the eigenvalue problem

(6.6) $F(\Phi,\lambda) = \Phi$

lead to periodic solutions $x(t;\Phi,\lambda)$. Furthermore, we have $F(0,\lambda) = 0$ for all $\lambda \in \mathbb{R}_+$.

We list the following results, which are known from the analysis of the equation (6.1):

(6.7) $(0, \frac{\pi}{2})$ is a bifurcation point of (6.6). This is a result of R. D. Nussbaum ([10]) using abstract fixed point theory. Nussbaum's result is a consequence of

(6.8) $i(F(.,\lambda),0) = 1$ for $\lambda < \frac{\pi}{2}$
$i(F(.,\lambda),0) = 0$ for $\lambda > \frac{\pi}{2}$.

(6.9) There exists a continuum C of nontrivial solutions of (6.6) which has the qualitative properties described in figure 3:

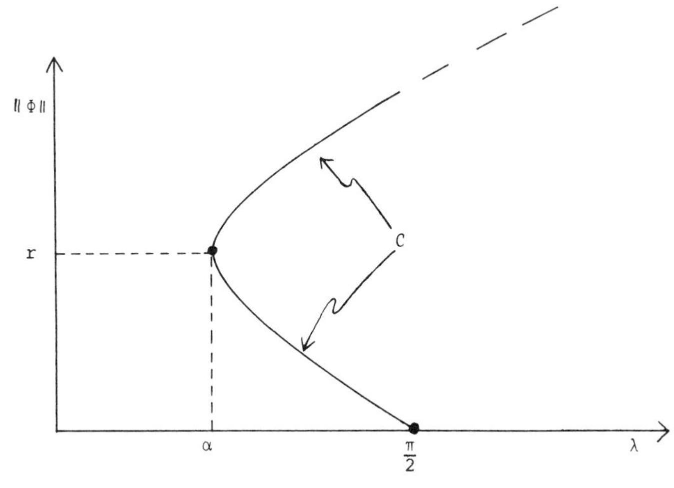

(6.9) is a result of H. O. Walther ([17]) using (6.7) and phase space arguments.

We describe our numerical approach to solve (6.6):

Using Euler differences, the original shift operator is replaced by an operator

(6.10) $F_\Delta : \mathbb{R}^n_* \times \mathbb{R}_+ \rightarrow \mathbb{R}^n_*$,

where $\mathbb{R}^n_*$ denotes the cone $\{x \in \mathbb{R}^n : \sigma = x_1 \leq x_2 \leq \ldots \leq x_n\} \subset \mathbb{R}^n$. Since $\mathbb{R}^n_*$ is homeomorphic to $\mathbb{R}^{n-1}_+$, (6.10) leads to an operator

(6.11) $F_\Delta : \mathbb{R}^n_+ \rightarrow \mathbb{R}^{n-1}_+$

According to (5.2), F_Δ induces a labeling $\ell^{F_\Delta} : \mathbb{R}^n_+ \rightarrow \{0,\ldots,n-1\}$.

Because of (6.8), theorem 6 is applicable. Indeed, an ℓ^{F_Δ}-Sperner-simplex σ^n which approximates the trivial solution (0,0) is easy to find for every triangulation τ of $\mathbb{R}^n_+$. A computer program was implemented to calculate $Ch(\sigma^n)$ with n=10 and meshτ = 0,01. The result was the following:

$Ch(\sigma^n)$ followed the line of trivial solutions until $(0, \frac{\pi}{2})$. At $(0, \frac{\pi}{2})$ $Ch(\sigma^n)$ branched off and followed the continuum of nontrivial solutions.

This numerical result is remarkable under several aspects:

Nothing is known about the differentiability properties of F. This is a serious obstacle for all algorithms which are for example of Newton type.

Our algorithm was the first to approximate the "unstable" branch of solutions (Φ,λ) with $\lambda < \frac{\pi}{2}$ and $\| \Phi \| < r$ (cf. figure 3).

Finally, our algorithm was able to pass the "critical points" $(0, \frac{\pi}{2})$ and (r,α) (cf. figure 3) without any difficulty.

REFERENCES

[1] Allgower, E. L., Georg, K., Simplicial and Continuation Methods for Approximating Fixed Points, to appear.

[2] Allgower, E. L., Keller, C. L., Reeves, T. E., A Program for the Numerical Approximation of a Fixed Point of an Arbitrary Continuous Mapping of the n-Cube or n-Simplex into Itself, Aerospace Research Laboratories Techn. Rep. 71-0257, Wright-Patterson Air Force Base, Ohio, November 1971.

[3] Brown, R. F., The Lefschetz Fixed Point Theorem, Scott, Foresman and Company, Glenview, Illinois, 1971.

[4] Cohen, D. I. A., On the Sperner Lemma, J. Comb. Theory 2 (1967), 585-587.

[5] Eaves, B. C., Homotopies for Computation of Fixed Points, Math. Programming 3 1 (1972), 1-22.

[6] Eaves, B. C., Saigal, R., Homotopies for Computation of Fixed Points on Unbounded Regions, Math. Programming 3 2 (1972), 225-237.

[7] Granas, M. A., Sur la méthode de continuité de Poincaré, C. R. Acad. Sc. Paris, t.282 (3 mai 1976), Série A, 983-985.

[8] Knaster, B., Kuratowski, C., Mazurkiewicz, S., Ein Beweis des Fixpunktsatzes für n-dimensionale Simplexe, Fund. Math. 14 (1929), 132-137.

[9] Krasnosel'skii, M. A., Topological Methods in the Theory of Nonlinear Integral Equations, Pergamon Press, Oxford 1964.

[10] Nussbaum, R. D., A Global Bifurcation Theorem with Applications to Functional Differential Equations, J. Funct. Anal. 19 (1975), 319-338.

[11] Prüfer, M., Sperner Simplices and the Topological Fixed Point Index, Sonderforschungsbereich 72, Universität Bonn, preprint no. 134.

[12] Rabinowitz, P. H., Some Global Results for Nonlinear Eigenvalue Problems, J. Funct. Anal. 7 (1971), 487-513.

[13] Scarf, H. H., The Approximation of Fixed Points of a Continuous Mapping, SIAM J. Appl. Math. 15 5 (1967), 1328-1343.

[14] Sperner, E., Neuer Beweis für die Invarianz der Dimensionszahl und des Gebietes, Abh. a. d. Math. Sem. d. U. Hamburg 6 (1928), 265-272.

[15] Todd, M. J., The Computation of Fixed Points and Applications, Springer Verlag, Berlin, Heidelberg, New York, 1976.

[16] Vertgeim, B. A., On an Approximate Determination of the Fixed Points of Continuous Mappings, Soviet Math. Dokl. 11 (1970), 295-298.

[17] Walther, H. O., Remarks on Periodic Solutions of Differential Delay Equations with Stable Linearization, to appear.

SELECTED CONTRIBUTIONS TO IMBEDDING METHODS FOR FINITE DIMENSIONAL PROBLEMS

J. W. Schmidt

Sektion Mathematik
Technische Universität Dresden
German Democratic Republic

1. INTRODUCTION

In recent time at the Technical University of Dresden some research work has been undertaken in applying the principle of imbedding or continuation to the numerical solution of nonlinear systems of equations

$$F(x) = 0, \quad F : R^N \to R^N. \tag{1}$$

The aim of the present paper is to give a survey of these results including a description of the developped algorithms as well as an interpretation of their theoretical and numerical behaviour. For proofs, however, it is referred to the

ISBN 0-12-729250-0

original papers.

Most of the classical methods for solving nonlinear equations will, in general, converge to a solution only if the initial approximations are sufficiently good. Besides descent methods imbedding methods have been employed with some success to overcome this disadvantage. They may be considered as a possibility to widen the domain of convergence, or, from another point of view, as a procedure to obtain sufficiently close starting points.

An important property for iterative algorithms is their implementability. This means in connection with imbedding methods that the algorithms can be carried out without using global constants such as Lipschitzian and regularity constants and that in a finite number of finite steps the original problem (1) belonging to the final parameter is reached. The present paper contains as its main part two examples for implementable imbedding methods, namely the method of D. Leder [4], [5], which reduces the imbedding problem to a minimizing problem and solves the latter one by a descent algorithm, and the method of H. Schwetlick [12], [13], which realizes the implementability by the new priciple of sufficiently weak increase.

The general idea of imbedding methods for computing a solution x^* of the equation (1) consists in considering a family of problems

$$H(x,t) = 0, \quad t \in [0,1], \tag{2}$$

defined by a homotopy or one-parameter imbedding function $H : R^N \times [0,1] \to R^N$ satisfying the two properties

$$H(x^o,0) = 0 \text{ for a given } x^o \in R^N \tag{3}$$

and

$$H(x,1) = F(x) \text{ for all } x \in R^N. \tag{4}$$

Therefore the homotopy path $x : [0,1] \to R^N$ defined by $H(x(t),t) = 0$ describes a path from the given vector $x(0) = x^o$ to the unknown solution $x(1) = x^*$. If in addition x and H are differentiable, the so-called Davidenko differential equation

$$\partial_1 H(x,t)x'(t) + \partial_2 H(x,t) = 0, \quad x(0) = x^o \tag{5}$$

is satisfied. Well-known examples for such homotopies are

$$H(x,t) = F(x) - (1-t)F(x^o) \tag{6}$$

(defect reducing homotopy)

and

$$H(x,t) = tF(x) + (1-t)(x-x^o) \tag{7}$$

(regularizising homotopy).

References: Ortega/Rheinboldt [21].

2. CONSTRUCTION OF LOCAL CONVERGENT METHODS USING THE DAVIDENKO EQUATION

In some earlier papers H. Kleinmichel [2], [3] investigated a general principle to construct iterative methods of higher order. Integrating the Davidenko differential equation by a quadrature formula of order k using the stepsize h = 1 he gets iterative procedures of the order k+1 for solving $F(x) = 0$. But because of the constant stepsize these methods possess in general only local convergence behaviour.

In detail, let the n-th approximation x^n to a zero x^* be already known. Then in the case of the defect reducing homotopy

$$H(x,t) = F(x)-(1-t)F(x^n)$$

one gets the Davidenko equation

$$x'(t) = -F'(x)^{-1}F(x^n), \quad x(0) = x^n, \tag{8}$$

and the numerical approximation for $x(1)$ is taken as x^{n+1}. As a first example the Taylor method applied to (8) with the stepsize h = 1 is mentioned:

$$x^{n+1} = x^n+x'(0) + \frac{1}{2!}x''(0) + \ldots + \frac{1}{k!}x^{(k)}(0),$$

where $x'(0), x''(0),\ldots,x^{(k)}(0)$ are successively computed by solving the linear equations

$$\{F'(x(t))x'(t)+F(x^n)\}_{t=0} = F'(x^n)x'(0)+F(x^n) = 0,$$

$$\frac{d}{dt}\{F'(x(t))x'(t)+F(x^n)\}_{t=0} = F'(x^n)x''(0)+F''(x^n)x'(0)^2 = 0,$$

$$\vdots \qquad\qquad \vdots$$

$$\frac{d^{k-1}}{dt^{k-1}}\{F'(x(t))x'(t)+F(x^n)\}_{t=0} = F'(x^n)x^{(k)}(0)+\ldots$$
$$\ldots+F^{(k)}(x^n)x'(0)^k = 0.$$

Obviously this method of the order k+1 requires the derivatives of F up to the k-th. This in general means a disadvantage. In further methods based on Runge Kutta formulas only first derivatives at different points appeared. For instance the classical Runge Kutta formula applied to (8) produces the following method of order 5:

$$F'(x_1^n)x_1^n+F(x^n) = 0$$
$$F'(x_1^n+0.5x_1^n)x_2^n+F(x^n) = 0$$
$$F'(x_1^n+0.5x_2^n)x_3^n+F(x^n) = 0$$
$$F'(x_1^n+x_3^n)x_4^n+F(x^n) = 0$$
$$x^{n+1} = x^n + \frac{1}{6}(x_1^n+2x_2^n+2x_3^n+x_4^n).$$

It is interesting that by combining the Newton method with its simplified version Traub developped a method of order k+1, which requires the computation of k function-values and of one value F' only, thus considerably reducing the amount per steps.

In the mentioned papers H. Kleinmichel also gives semi-local convergence theorems of the Kantorovich-type for these methods.

Further references: Ortega/Rheinboldt [21], L. Bittner [1].

3. D. LEDER'S IMPLEMENTABLE IMBEDDING METHOD

An important approach in obtaining a zero $x^* = x(1)$ of F by imbedding methods is the following. A partition

$$0 = t_0 < t_1 < \cdots < t_M = 1$$

of the interval [0,1] is chosen, and one proceeds solving successively the chain of subproblems

$$H(x^{n+1}, t_{n+1}) = 0, \quad n=0,\ldots,M-1$$

by some iterative method, which uses the accepted solution x^n for $t = t_n$ as starting value for computing x^{n+1}. If the stepsizes $h_n = t_{n+1}-t_n$ are sufficiently small, then, hopefully, in solving the subproblems convergence will occur. On the other side, in order to arrive at the final parameter t=1 and to minimize the total amount of work for obtaining x^*, the stepsizes should be chosen as large as possible.

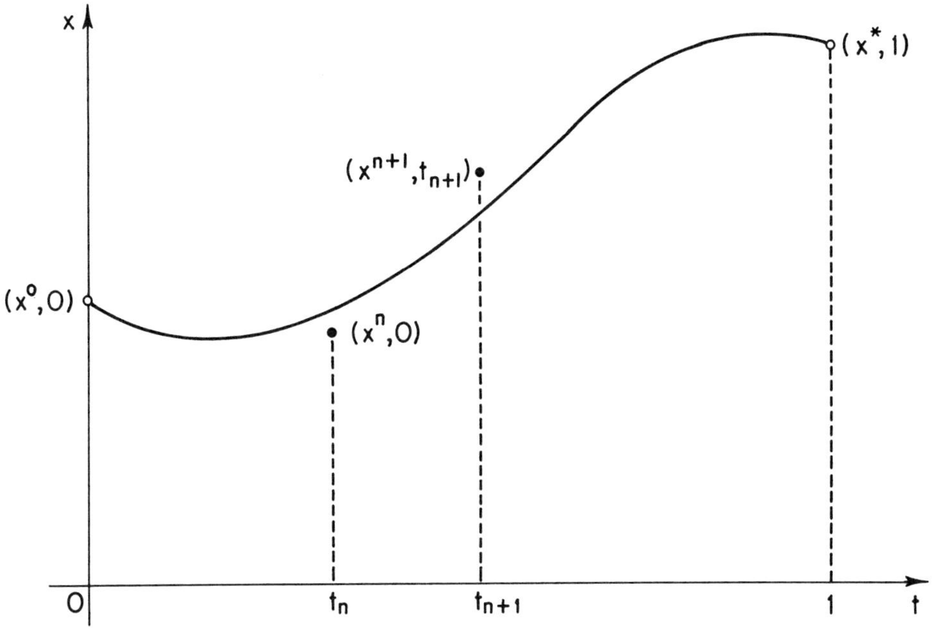

FIGURE 1. Finite imbedding algorithm in regular cases.

Several theoretical estimates for the choice of the step-sizes have been elaborated by means of global informations on the homotopy H. In general, these results are not immediately applicable. The apparantly first implementable algorithm based on imbedding is due to D. Leder [4], [5]. He developped an algorithm with adaptive stepsize control without using global constants. The main idea of his method will be presented now. At first the problem (2) is equivalently transformed to

$$G(x,t) = \begin{bmatrix} H(x,t) \\ m(1-t) \end{bmatrix} = 0,\ m \neq 0\ ,\ m \text{ constant} \tag{9}$$

and, in a second step, to

$$g(x,t) = \frac{1}{2} \|G(x,t)\|^2 = \frac{1}{2}(\|H(x,t)\|^2 + m^2(1-t)^2) = \min! \tag{10}$$

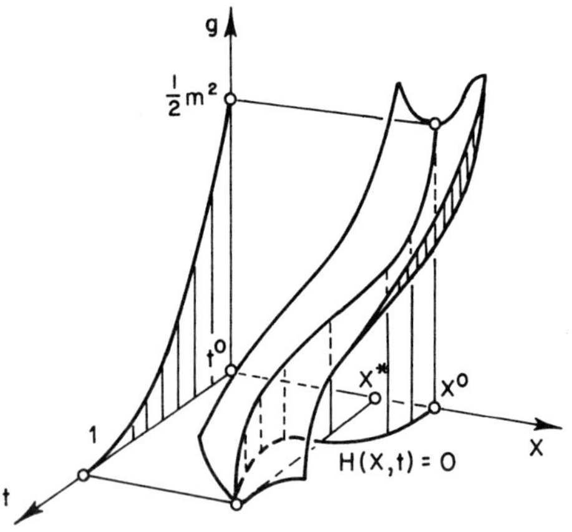

FIGURE 2. The functional g.

The last problem is now solved for instance by the damped Newton method. It is well-known that the Newton direction

$$p(x,t) := \begin{bmatrix} s(x,t) \\ \tau(x,t) \end{bmatrix} := G'(x,t)^{-1}G(x,t) =$$

$$= \begin{bmatrix} \partial_1 H(x,t)^{-1}\{H(x,t)+(1-t)\partial_2 H(x,t)\} \\ t-1 \end{bmatrix}$$

is a descent direction with respect to g:

$$g'(x,t)^T p(x,t) = \|G(x,t)\|^2 > 0 \text{ for } (x,t) \neq (x^*,1). \quad (11)$$

Thus for small $\gamma > 0$ the monotonicity condition

$$g(x-\gamma s(x,t), t-\gamma\tau(x,t)) < g(x,t) \quad (12)$$

is satisfied. In the following algorithm the condition (12) is sharpened in the sense of the Goldstein-Armijo test to assure convergence.

ALGORITHM (D. LEDER): Let $\alpha, q \in (0,1)$, $x^0 \in R^N$ with $H(x^0,0) = 0$ be given and set $t_0 = 0$. Then in the n-th step compute $s^n \in R^N$ by solving the linear system

$$H(x^n,t_n)+(1-t_n)\partial_2 H(x^n,t_n) = \partial_1 H(x^n,t_n)s^n$$

and define

$$\tau_n = t_n - 1,$$

choose the damping parameter as the largest number

$$\gamma_n \in \{1,\alpha,\alpha^2,\ldots\} \tag{13}$$

with the property

$$\|G(x^n-\gamma_n s^n, t_n-\gamma_n\tau_n)\|^2 \le (1-\gamma_n q)\|G(x^n,t_n)\|^2 \tag{14}$$

and set at the end of the n-th step

$$x^{n+1} = x^n-\gamma_n s^n, t_{n+1} = t_n-\gamma_n\tau_n.$$

In general, the computing time for performing the search procedure (13), (14) is reduced if γ_n is the greatest number according to (14) and

$$\frac{\gamma_n}{c_n} \in \{1,\alpha,\alpha^2,\ldots\},\ c_n = \min(1,\gamma_{n-1}/\alpha). \tag{15}$$

The convergence behaviour of the described algorithm in the case of a regular homotopy (16) is characterized by the following

THEOREM: Suppose that the level set

$$W_0 = \{\binom{x}{t} \in R^{N+1} : \|G(x,t)\| \le \|G(x^0,0)\|\}$$

is compact, the homotopy H is twice continuously differentiable on W_0 and the matrix $\partial_1 H(x,t)$ is regular on W_0 (then there exists a constant K with

$$\|\partial_1 H(x,t)^{-1}\| \leq K \text{ for } (x,t) \in W_o). \tag{16}$$

Under these assumptions Leder's algorithm is well-defined, especially the γ_n-search is finite in every step, and a sequence (x^n) is generated converging to a zero x^* of F. Moreover, there exists an index n_o with $\gamma_n = 1$ for all $n \geq n_o$, i.e. for $n \geq n_o$ the sequence (x^n) is defined by the undamped Newton method applied to the equation (1),

$$x^{n+1} = x^n - s^n,\ F(x^n) = F'(x^n)s^n \text{ for } n \geq n_o,$$

and the convergence order is two.

Leder [4], [5] has also given derivative-free variants based on the concept of consistent approximations. In general, however, the descent property (12) is now lost except the consistency parameters are sufficiently small.

The choice of the parameter m in (9) is very crucial. If m is too small a large number of steps is needed to approach x^*. If m is too large the numerical approximation of the homotopy path is bad, which leads possibly to an earlier breakdown. This is demonstrated by the following example [5], which uses the defect reducing homotopy (6).

Example (Rosenbrock):

$$F_1(x) = 1 - x_1$$

$$F_i(x) = 10(i-1)(x_i - x_{i-1})^2,\ i=2,\ldots,N$$

Starting vector: $x^o = (-1.2,\ldots,-1.2,-1)^T$

Solution: $x^* = (1,\ldots,1,1)^T$

N	m	number of steps total	damped	number of F-Values	defect at the last step
2	4.92	65	63	259	2.96-9
2	25.0	4	1	14	0.00-10
5	10.0	72	69	502	3.64-10
5	25.0	38	35	261	1.60-9

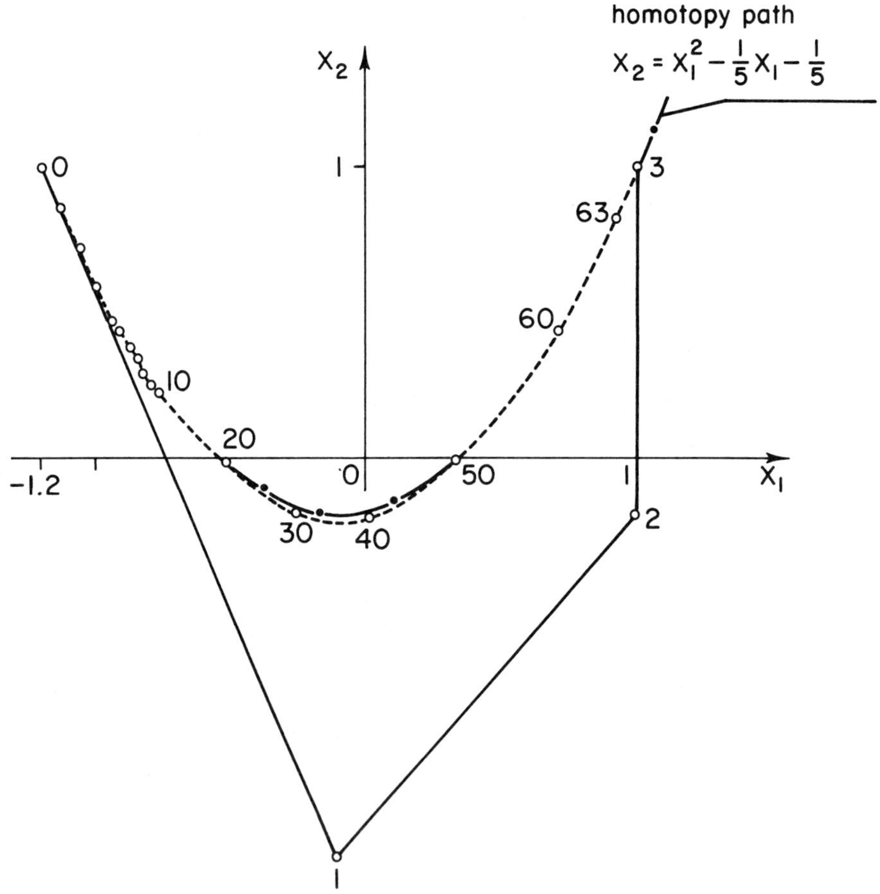

FIGURE 3. Numerical homotopy pathes in Rosenbrock's example (m=4.92 ------, m=25.0 ———)

Further references: Ortega/Rheinboldt [21], Avila [14], Deuflhard [17], [18], Feilmeier [19], Laasonen [20], Rheinboldt [22], H. Schwetlick [11], Wacker [25], [26].

4. IMPLEMENTABLE IMBEDDING METHODS BASED ON SCHWETLICK'S PRINCIPLE OF SUFFICIENTLY WEAK INCREASE (REGULAR CASE)

In connection with imbedding methods H. Schwetlick [12], [13] introduced the new principle of sufficiently weak increase. Let $\varepsilon > 0$ be given and (x^n, t_n) already be known. Then (x^{n+1}, t_{n+1}) is accepted as an appropriate successor of (x^n, t_n) if

$$\|H(x^{n+1}, t_{n+1})\| \leq \|H(x^n, t_n)\| + \varepsilon(t_{n+1} - t_n). \tag{17}$$

Thus, if the final parameter $t_M = 1$ is reached, it is immediately seen that x^M is an ε-approximation to a zero of F:

$$\begin{aligned} \|F(x^M)\| = \|H(x^M, t_M)\| &\leq \|H(x^o, t_o)\| + \\ &+ \varepsilon(t_M - t_{M-1}) + \dots + \varepsilon(t_1 - t_o) = \varepsilon. \end{aligned}$$

Assuring (17) methods of the type

$$x^{n+1} = x^n - h_n p(x^n, t_n, h_n)$$

are considered, where $p : R^N \times [0,1] \times [0,1] \to R^N$ is the so-called increment function. Because of

$$H(x-hp,t+h) = H(x,t) + \\ + h\{-\partial_1 H(x,t)p + \partial_2 H(x,t)\} + o(h) \tag{18}$$

the requirement (17) is satisfied in the case of sufficiently small h if the increment function is consistent to the homotopy H in the sense of

$$p(x,t,0) = \partial_1 H(x,t)^{-1}\partial_2 H(x,t) \text{ for } (x,t) \in R^N \times [0,1]. \tag{19}$$

Incorporating the Goldstein-Armijo search procedure the following algorithm is suggested.

BASIC ALGORITHM (H. SCHWETLICK)

Let the prescribed accuracy $\varepsilon > 0$, the reducing factor $\alpha \in (0,1)$, $h_0 \in (0,1]$, the initial vector $x^0 \in R^N$ with $H(x^0,0) = 0$ be given and set $t_0 = 0$. Then in the n-th step choose the steplength h_n as the largest number such that

$$\frac{h_n}{c_n} \in \{1,\alpha,\alpha^2,\dots\} \tag{20}$$

and

$$\|H(x^n-h_n p(x^n,t_n,h_n),t_n+h_n)\| \le \|H(x^n,t_n)\| + \varepsilon h_n, \tag{21}$$

where $c_n = 1-t_n$ or, more advantageously, $c_n = \min(1-t_n,h_{n-1}/\alpha)$. Finally define

$$x^{n+1} = x^n-h_n p(x^n,t_n,h_n), \quad t_{n+1} = t_n+h_n. \tag{22}$$

Because of $h_n \le c_n \le 1-t_n$ in both cases of the definition of c_n the iteration does not leave the interval [0,1] of the homotopy parameter. The finiteness properties of the basic algorithm in the case of a regular homotopy are summarized by the following

THEOREM (H. SCHWETLICK)

Let the homotopy H be continuously differentiable and regular on $R^N \times [0,1]$ in the sense of (16), the increment function p be consistent with respect to H and $\varepsilon > 0$ be given. Then the basic algorithm is well-defined, especially in every step an appropriate successor is found in a finite number of search steps (2o), (21) and the algorithm terminates after a finite number of steps with $t_M=1$ and the related x^M satisfies $\|F(x^M)\| \le \varepsilon$.

Some examples of consistent increment functions including a definition of an order concept will be given now. With respect to the equation (18) in the paper [9] an increment function is said to be of order $k \ge 1$ if for $(x,t) \in R^N \times [0,1]$ and $h \to 0$

$$H(x-hp(x,t,h),t+h)-H(x,t) = O(h^k). \qquad (23)$$

Let $t = t_n \in [0,1]$, $x = x^n \in R^N$ be already known and $h > 0$, too. Then the vectors $y = y(x,t,h) = x-hp(x,t,h)$ satisfying approximatively the equation

$$H(y,t+h)-H(x,t) = 0 \tag{24}$$

are suitable for determining increment functions p. At first Taylor expansions for the solution y of (24) are mentioned.

For instance, by the linear Taylor formula

$$y(x,t,h) \approx y(x,t,0) + hy'(x,t,0)$$

because of $y(x,t,0) = x$ and $y'(x,t,0) = -\partial_1 H(x,t)^{-1}\partial_2 H(x,t)$ the simple Euler-Cauchy increment function of order 1

$$p(x,t,h) = \partial_1 H(x,t)^{-1}\partial_2 H(x,t) \tag{25}$$

is obtained and, analogously, second order Taylor approximation leads to the increment function of order 2

$$\begin{aligned} p(x,t,h) &= \partial_1 H(x,t)^{-1}\partial_2 H(x,t) + \\ &+ \frac{h}{2}\,\partial_1 H(x,t)^{-1}\{\partial_{11}H(x,t)(\partial_1 H(x,t)^{-1}\partial_2 H(x,t))^2 \\ &- 2\partial_{12}H(x,t)\partial_1 H(x,t)^{-1}\partial_2 H(x,t)+\partial_{22}H(x,t)\}. \end{aligned} \tag{26}$$

Secondly, a more interesting class of increment functions is based on the PC-principle, where the corrector is constructed by solving the equation (24) approximatively by a finite number of Newton steps. Starting, for instance, with the predictor

$$v^0 = x \text{ or } v^0 = x-h\partial_1 H(x,t)^{-1}\partial_2 H(x,t) \tag{27}$$

the corrector is now of the form

$$v^{i+1} = v^i - \partial_1 H(v^i, t+h)^{-1} \{H(v^i, t+h) - H(x,t)\}, \tag{28}$$
$$i=1,\ldots,m-1.$$

If the integer m is fixed, the increment function

$$p(x,t,h) = \frac{1}{h} \{x - v^m\}$$

corresponding to (27) and (28) is of the order 2^{m-1} or 2^m, respectively. But, in general, an adaptive choice of m is more appropriate than a fixed one. Let be

$$Q_i = \frac{1}{h} \{\|H(v^i, t+h)\| - \|H(x,t)\|\}.$$

Then if $Q_i \leq \varepsilon$, the inequality (21) of the basic algorithm holds and $x^{n+1} = v^i$ is accepted as an approximation at $t_{n+1} = t+h$. Otherwise as long as $Q_i \leq \mu Q_{i-1}$ with some $\mu \in (0,1)$ Newton's iteration (28) is continued with the stepsize h. If $Q_i > \mu Q_{i-1}$ in accordance with the basic algorithm the stepsize h is reduced multiplying its current value by α and one proceeds by going to (27).

In [9] there are also given increment functions based on Runge-Kutta methods and other ODE-solvers.

For numerical examples see [12].

Finally it is mentioned that because of a large number of subproblems in the homotopy chain it is, in general, not advantageous to choose the accuracy ε very small. Starting, for instance, with the relatively large value $\varepsilon = 10^{-2}$ the final vector x^M is possibly a suitable initial vector for the undamped Newton method applied to the given problem $F(x) = 0$. If not, the homotopy is changed by substituting x^0 by the last x^M and $\varepsilon > 0$ is reduced until the new final vactor x^M produces a convergent Newton sequence. In [12] this combination is called a composed algorithm.

Further references: See the references of chapter 3.

5. IMPLEMENTABLE IMBEDDING ALGORITHMS IN THE CASE OF WEAKLY SINGULAR HOMOTOPIES

The methods discussed so far will, in general, fail if the regularity of the homotopy is not assured. Problems arising then have been treated by Haselgrove, Anselone/Moore, Davis and Meyer in some earlier papers. In recent time in the case of weakly singular homotopies an implementable imbedding algorithm is proposed by R. Menzel [6], [7] extending Schwetlick's principle of sufficiently weak increase and, in addition, using general ideas of the non-implementable algorithms for instance of Haselgrove and R. Menzel/H. Schwetlick [8].

In detail, defining $G : R^{N+1} \to R^N$ by the homotopy H according to

$$G(u) = H(x,t), u = (x,t) \in R^N \times R^1, \tag{30}$$

it is assumed that there exists a twice continuously differentiable curve $w : [0,1] \to R^{N+1}$ which connects the given value $w(0) = (x^0, 0)$ with the unknown value $w(1) = (x^*, 1)$ where $F(x^*) = 0$ and which satisfies the homotopy equation

$$G(w(s)) = 0 \text{ for } s \in [0,1]. \tag{31}$$

Notice that now the curve parameter s is in general different from the homotopy parameter t.

Moreover let G be twice continuously differentiable in a neighbourhood of the curve w and let G' be of full rank:

$$\text{rk } G'(w(s)) = \text{rk } [\partial_1 H(x(s),t(s)), \partial_2 H(x(s),t(s))] = N$$
$$\text{for } s \in [0,1]. \tag{32}$$

In this case the homotopy is said to be weakly singular.

Finally let the conditions

$$\dot{w}(s) \neq 0 \text{ for } s \in [0,1], \ \dot{w}(0)^T e^{N+1} > 0, \tag{33}$$

$$w(s)^T e^{N+1} < 1 \text{ for } s \in [0,1)$$

and

$$\text{rk } F'(x^*) = N \tag{34}$$

be valid.

In the description of the algorithm suggested by R. Menzel [6], [7] the last computed approximation is denoted by $u^n = (x^n, t_n) \in R^{N+1}$ and supposed to be in a sufficiently small neighbourhood of the curve w. Then, because (32) holds, the tangent direction $v^n \in R^{N+1}$ is apart from sign uniquely determined by

$$G'(u^n)v^n = 0, \ \|v^n\| = 1. \tag{35}$$

The sign may be choosen such that $(v^n)^T v^{n-1} \geq 0$ (and $(v^o)^T e^{N+1} \geq 0$). Moreover, in a neighbourhood of u^n and for sufficiently small $|\tau|$ the system of equations

$$G(u) = G(u^n), \ (u-u^n)^T v^n = \tau \tag{36}$$

is uniquely solvable by $u = w^n(\tau)$, and with a constant $\gamma > 0$ holds

$$\|w^n(\tau) - (u^n + \tau v^n)\| \leq \gamma\tau^2. \tag{37}$$

The function $w^n : (-\varepsilon_n, \varepsilon_n) \rightarrow R^{N+1}$, $\varepsilon_n > 0$, obviously is a local parametrisation of the manifold of solutions of (36).

A continuous function $p : R^{N+1} \times R^1 \rightarrow R^{N+1}$ is called a consistent generalized increment function if

$$p(u,0) \neq 0 \text{ for all } u \text{ on the curve } w \tag{38}$$

and

$$\lim_{\tau \to 0} \left\| \frac{p(u,\tau)}{\|p(u,\tau)\|} - v \right\| = 0 \tag{39}$$

for all u in a neighbourhood of the curve w. Here v is the suitably turned tangent direction in u determined by $G'(u)v = 0$, $\|v\| = 1$. Obviously $p(u,\tau) = v$ is a consistent increment function.

Now $u^{n+1} = (x^{n+1}, t_{n+1}) \in R^{N+1}$ where

$$u^{n+1} = u^n + \tau_n p(u_n, \tau_n) \tag{40}$$

is accepted as a successor of $u^n = (x^n, t_n)$ if the inequality of sufficiently weak increase

$$\|G(u^{n+1})\| \leq \|G(u^n)\| + \tau_n \varepsilon \tag{41}$$

is satisfied with a given $\varepsilon > 0$. Otherwise τ_n is reduced multiplying its current value by an $\alpha \in (0,1)$ until (41) holds.

In this algorithm two cases can occur. At first let the tangent direction be turned forward, $(v^n)^T e^{N+1} > 0$. Then the intersection point z^n of the tangent $z = u^n + \tau v^n$ with the hyperplane

$$\sigma = \{z \in R^{N+1} : z^T e^{N+1} = 1\}$$

is of importance. One gets

$$z^n = z(\bar{\tau}_n), \quad \bar{\tau}_n = \frac{1-t_n}{(v^n)^T e^{N+1}}. \tag{42}$$

If setting

$$u^{n+1} = u^n + \bar{\tau}_n v^n \tag{43}$$

inequality (41) is fulfilled with $\tau_n = \bar{\tau}_n$, the algorithm can be stopped.

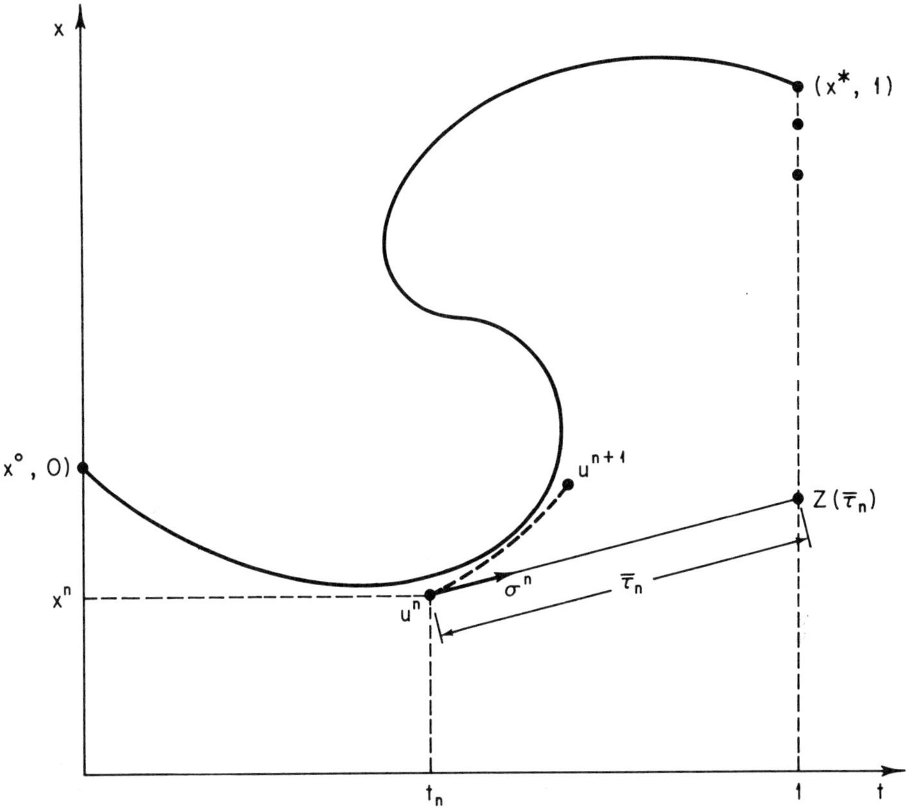

FIGURE 4. Weakly singular homotopy - the cases (40) and (42) of the implementable algorithm.

Otherwise it is possible to add Newton's method

$$y^0 = ((u^n+\bar{\tau}_n v^n)^T e^1, \ldots, (u^n+\bar{\tau}_n v^n)^T e^N)^T$$

$$F(y^i) + F'(y^i)(y^{i+1}-y^i) = 0, \quad i=0,1,\ldots \tag{44}$$

applied to the initial problem $F(x) = 0$, which may be of success in the sense of

$$\|F(y^{i+1})\| \le \mu \|F(y^i)\|, \quad \mu \in (0,1) \text{ constant.} \tag{45}$$

Then there exists an index $i = i_o$ such that (41) holds now meaning

$$\|H(y^{i_o},1)\| \le \|G(u^n)\| + \varepsilon\bar{\tau}_n, \tag{46}$$

and the procedure is stopped, too.

If for $(v^n)^T e^{N+1} > 0$ this way is not successful, the algorithm proceeds as in the following second case, i.e. the tangent direction is turned backward, $(v^n)^T e^{N+1} \le 0$. Then the algorithm is continued as described by (40) and (41) where in addition $\tau_n \le \bar{\tau}$ is required with a given $\bar{\tau} > 0$.

Assuming the homotopy G, the curve w and the increment function p are of the kind mentioned above the properties of the algorithm are reflected by the following

THEOREM (R. MENZEL)

If $\varepsilon > 0$ and $\bar{\tau} > 0$ are sufficiently small, the algorithm is well-defined, especially in every step only a finite number of trials (41) is necessary, and the algorithm terminates after a finite number of steps with a vector $u^M = (x^M, t_M)$

where $t_M = 1$ and $\| F(x^M) \| \leq c\varepsilon$. Here $c > 0$ is a constant unfortunately, in contrast to the regular case, not known a priori.

In [7] some examples of generalized increment functions are given, among them one is based on the PC-principle. Let be $\tau \in R^1$, $u = u^n \in R^{N+1}$, $v = v^n$ the corresponding tangent direction and m an integer. Then, starting with the predictor

$$z^o = u + \tau v \tag{47}$$

the corrector is constructed solving the system

$$G(z) = G(u), \; v^T(z-u) = \tau \tag{48}$$

by Newton's method. That means

$$\begin{aligned} &G(z^i)+G'(z^i)(z^{i+1}-z^i) = 0 \\ &v^T(z^{i+1}-z^i) = 0, \; i=0,\dots,m-1. \end{aligned} \tag{49}$$

The related increment function

$$p(u,\tau) = \frac{1}{\tau}\{z^m-u\} \tag{50}$$

is consistent and of the order 2^m.

The non-implementable imbedding algorithm analyzed by R. Menzel/H. Schwetlick [8] is closely connected with the increment function (50). There, however, $G(u^n) = 0$ is supposed and the Newton method (49) is continued as long as

$$\|G(z^{i+1})\| \leq \mu\|G(z^{i})\|, \quad (z^{i+1})^{T}e^{N+1} \leq 1 \tag{51}$$

where $\mu \in (0,1)$ constant. Otherwise the stepsize τ_n is reduced. If (49) is always successful in the sense of (51), the next approximation is defined by

$$u^{n+1} = \lim_{i\to\infty} z^{i}. \tag{52}$$

In [8] for this modification it is shown that in a finite number of steps each of them now requiring an infinite process the final parameter $t_M = 1$ is reached such that the corresponding vector x^M of the predictor taken as starting vector produces a quadratically convergent Newton sequence for the initial problem $F(x) = 0$.

The following numerical example is due to [8].

Example (Freudenstein/Roth):

$$F_1(x) = x_1 + 5x_2^2 - x_2^3 - 2x_2 - 13$$

$$F_2(x) = x_1 + x_2^2 - x_2^3 - 14x_2 - 29.$$

Starting vector: $x^0 = (15.0, -2.0)^T$

Solution: $x^* = (5.0, 4.0)^T$.

Notice that the Jacobian is singular along the lines

$$x_2 = (2 \pm \sqrt{22})/3.$$

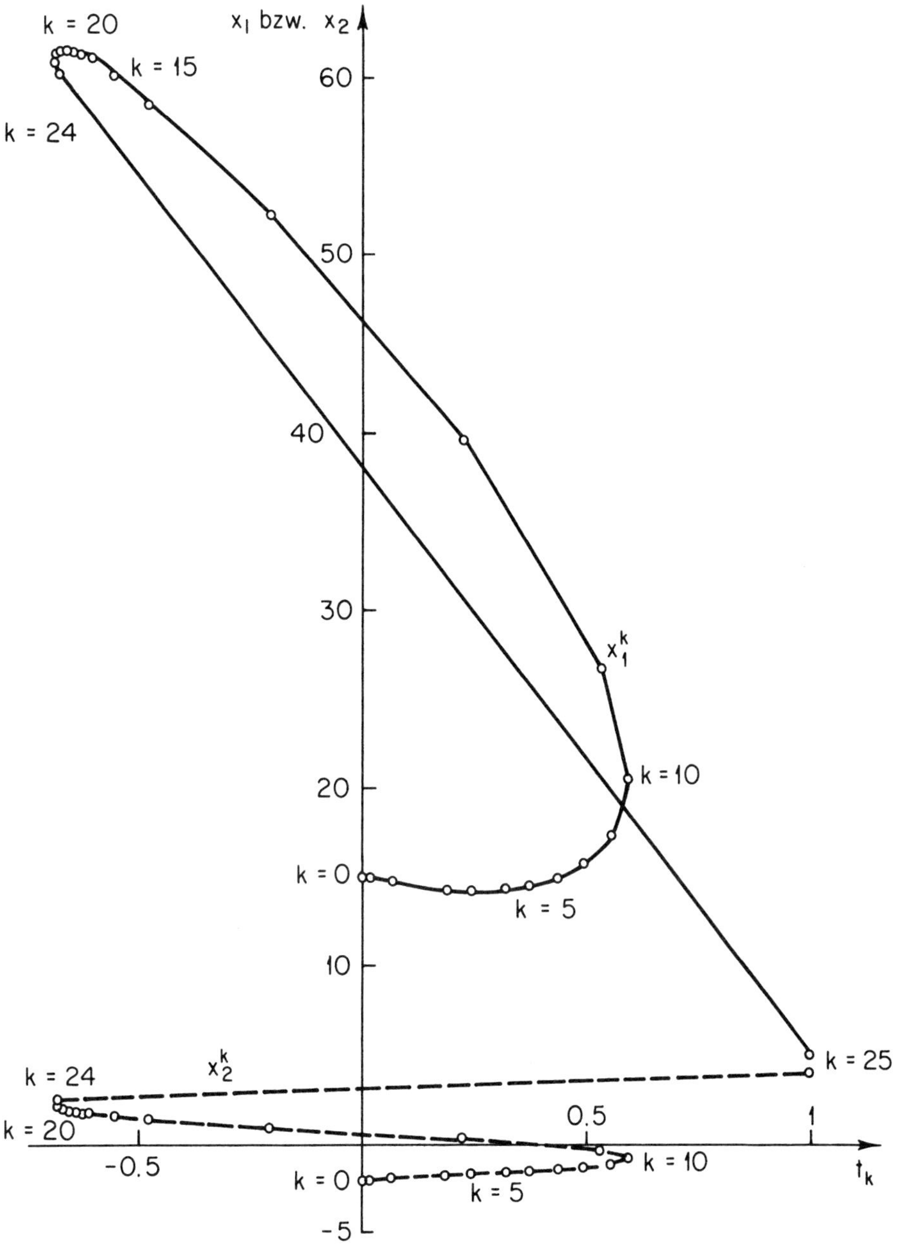

FIGURE 5. Numerical homotopy path in the Freudenstein/ Roth example.

Further references: Avila [14], Branin [15], Chua/ Ushida [16], Deuflhard [17], [18], Riks [23].

6. A LOCAL CONVERGENT METHOD FOR THE DETERMINATION OF TURNING POINTS

In the case of weakly singular homotopies turning points may be occur and also in some other problems one is interested just in these points. It is supposed that the general assumptions of chapter 5 are satisfied, especially condition (32). A point $\bar{u} = (\bar{x},\bar{t}) \in R^{N+1}$ is called a (right sided) turning point of G if

$$(u-\bar{u})^T e^{N+1} \leq 0 \qquad (53)$$

for all u in a neighbourhood of $\bar{u}$, for which $G(u) = 0$ holds.

Following a paper of G. Pönisch/H. Schwetlick [10] let u^n, u^{n-1} be two sufficiently good approximations to $\bar{u}$ and v^n, v^{n-1} the corresponding normed tangent directions. Then, in accordance with (36), there exists a local parametrisation $w^n : (-\varepsilon_n,\varepsilon_n) \to R^{N+1}$, $\varepsilon_n > 0$ solving the system

$$G(u) = 0, (u-u^n)^T v^n = \tau \qquad (54)$$

and therefore $G(w^n(\tau)) = 0$, $(w^n(\tau)-u^n)^T v^n = \tau$ for $\tau \in (-\varepsilon_n,\varepsilon_n)$

is valid. Hence, because of (53) the problem is transformed into the local optimization problem

$$g_n(\tau) := w^n(\tau)^T e^{N+1} \rightarrow \max! \tag{55}$$

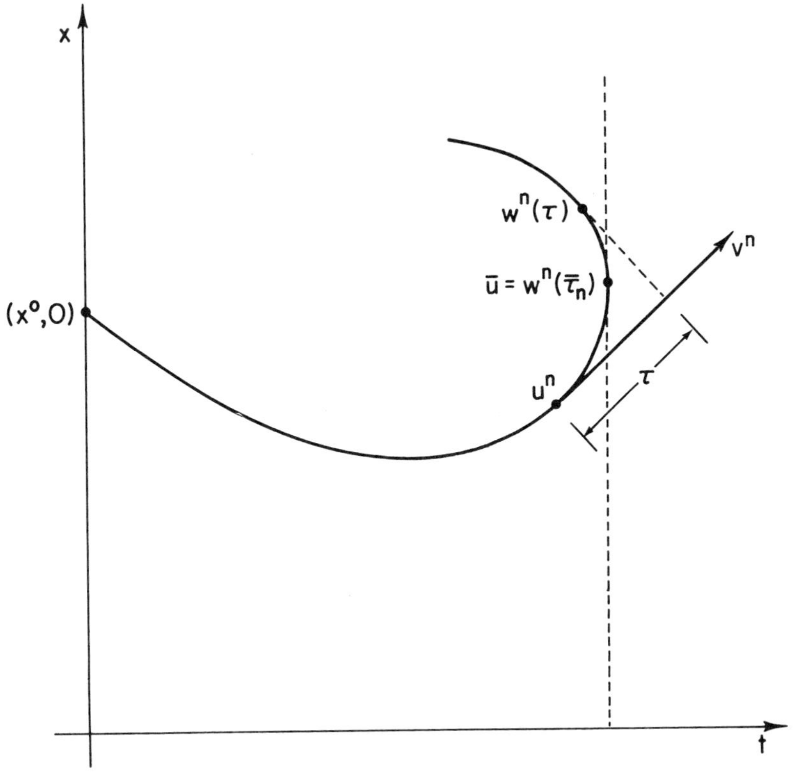

FIGURE 6. Turning point and local parametrisation.

Now, in the first part problem (55) is solved approximatively by Hermite interpolation. With $\sigma = (u^{n-1}-u^n)^T v^n$ one gets the relations

$$w^n(0) = u^n, \ \dot{w}^n(0) = v^n$$
$$w^n(\sigma) = u^{n-1}, \ \dot{w}^n(\sigma) = v^{n-1}/(v^{n-1})^T v^n. \tag{56}$$

Thus the values $g_n(0)$, $g_n'(0)$, $g_n(\sigma)$, $g_n'(\sigma)$ are known. Let $\tilde{\tau}_n$ be the maximizer of the corresponding cubic Hermite polynomial. After the calculation of $\tilde{\tau}_n$ in a second part u^{n+1} is computed by applying Newton's method to

$$G(u) = 0, \ (u-u^n)^T v^n = \tilde{\tau}_n \tag{57}$$

where the starting vector is choosen as $u^n + \tilde{\tau}_n v^n$.

In [10] the following properties are proved. If G is sufficiently smooth in a neighbourhood of a simple turning point $\bar{u}$ and if the starting vectors u^0, u^1 besides $G(u^0) = G(u^1) = 0$ are sufficiently close to $\bar{u}$, then the algorithm is well-defined and the corresponding sequence (u^n) converges quadratically to $\bar{u}$.

Further references: Simpson [24]; see also the references of chapter 5.

REFERENCES

For papers appeared before 1970 it is, in general, referred to the monograph [21].

[1] Bittner, L., Einige kontinuierliche Analogien von Iterationsverfahren, ISNM 7 (1967), 114-135.

[2] Kleinmichel, H., Existenzsätze und numerische Verfahren für nichtlineare Gleichungen in Banachräumen, Diss., TU Dresden 1967.

[3] Kleinmichel, H., Stetige Analoga und Iterationsverfahren für nichtlineare Gleichungen in Banachräumen, Math. Nachr. 37 (1968), 313-343.

[4] Leder, D., Automatische Schrittweitensteuerung bei global konvergenten Einbettungsverfahren, Z. Angew. Math. Mech. 54 (1974), 319-324.

[5] Leder, D., Zur Lösung nichtlinearer Gleichungssysteme mittels diskreter Einbettungsmethoden, Diss. A , TU Dresden 1975.

[6] Menzel, R., Über die Lösung nichtlinearer Gleichungssysteme bei regulärer und schwach singulärer Einbettung, Diss.A , TU Dresden 1978.

[7] Menzel, R., Ein implementierbarer Algorithmus zur Lösung nichtlinearer Gleichungssysteme bei schwach singulärer Einbettung, Beitr. Numer. Math. (eingereicht).

[8] Menzel, R., Schwetlick, H., Zur Lösung parameterabhängiger nichtlinearer Gleichungen mit singulären Jacobi-Matrizen, Inform. TU Dresden 07-11-75, 07-12-75 und Numer. Math. (im Druck).

[9] Menzel, R., Schwetlick, H., Über einen Ordnungsbegriff bei Einbettungsalgorithmen zur Lösung nichtlinearer Gleichungen, Computing 16 (1976), 187-199.

[10] Pönisch, G., Schwetlick, H., Ein lokal überlinear konvergentes Verfahren zur Bestimmung von Rückkehrpunkten implizit definierter Raumkurven, Inform. TU Dresden 07-30-77 und Numer. Math. (eingereicht).

[11] Schwetlick, H., Über die Realisierung und Konvergenz von Mehrschrittverfahren zur iterativen Lösung nichtlinearer Gleichungen, Z. Angew. Math. Mech. 54 (1974), 479-493.

[12] Schwetlick, H., Ein neues Prinzip zur Konstruktion implementierbarer, global konvergenter Einbettungsalgorithmen, Beitr. Numer. Math. 4 (1975), 215-228 und 5 (1976), 201-206.

[13] Schwetlick, H., Über die numerische Lösung nichtlinearer Probleme, Diss.B, TU Dresden 1976.

[14] Avila, J. H., The feasibility of continuation methods for nonlinear equations, SIAM J. Numer. Anal. 11 (1974), 102-122.

[15] Branin, F. J. jr., Widely convergent method for finding multiple solutions of simultaneous nonlinear equations, IBM J. Res. Develop. 16 (1972), 504-522.

[16] Chua, L. O., Ushida, A., A switching-parameter algorithm for finding multiple solutions of nonlinear resistive circuits, Circuit Theory and Appl. 4 (1976), 215-239.

[17] Deuflhard, P., A modified Newton method for the solution of ill-conditioned systems ... Numer. Math. 22 (1974), 289-315.

[18] Deuflhard, P., A stepsize control for continuation methods with special application to multiple shooting techniques, TU Munich Report Math. 76-07.

[19] Feilmeier, M., Numerische Aspekte bei der Einbettung nichtlinearer Probleme, Computing 9 (1972), 355-364.

[20] Laasonen, P., An imbedding method of iteration with global convergence, Computing 5 (1970), 253-258.

[21] Ortega, J. M., Rheinboldt, W. C., Iterative solution of nonlinear equations in several variables, Academic Press, New York - London (1970).

[22] Rheinboldt, W. C., An adaptive continuation process for solving systems of nonlinear equations, Univ. Maryland, Comp. Sc. Dep. TR-393-75.

[23] Riks, E., The application of Newton's method to the problem of elastic stability, J. Appl. Mech. 39 (1972), 1060-1065.

[24] Simpson, R. B., A method for the numerical determination of bifurcation states ... SIAM J. Numer. Anal. 12 (1975), 439-451.

[25] Wacker, H. J., Ein Iterationsverfahren zur Lösung spezieller nichtlinearer Randwertprobleme, Computing 9 (1972), 275-291.

[26] Wacker, H. J., Minimierung des Rechenaufwandes bei Globalisierungen spezieller Iterationsverfahren vom Typ Minimales Residuum, Computing 18 (1977), 209-224.

OPTIMAL STEPSIZE CONTROL FOR THE GLOBALIZED NEWTON METHOD

Hansjörg Wacker
E. Zarzer
W. Zulehner

Institut für Mathematik
Universität Linz
Linz, Austria

I. INTRODUCTION

The nonlinear operator equation $T(x) = 0$, $T: D \subset X \to Y$, X,Y Banach-spaces is to be solved using an imbedding method $T(x,s) = 0$. We do this by solving a finite sequence $T(x,s_n)=0$ (with $0 = s_o < s_1 < \ldots < s_N = 1$) where we use the Newton method (N.V.) or the modified Newton method (M.N.V.) to compute the solution $x(s_{n+1})$ of $T(x,s_{n+1})=0$ by means of the solution $x(s_n)$

ISBN 0-12-729250-0

of $T(x,s_n)=0$. (Sufficient conditions for the validity of this procedure can be found e.g. in Avila [1] , Ficken [7] , Wacker [16]) .

The following problems, which have not yet been solved completely up to now, require attention:

(1) Selection of a problem-adapted homotopy $T(x,s)$: The known results are of the type that for a chosen homotopy some sufficient conditions are to be checked. Only a few papers exist ([6] , [16]) concerning the selection of an imbedding closely related to the problem and with minimal computational effort - leaving aside problems (e.g. physical problems) for which one has to choose a special type of imbedding because of the nature of the equation.

(2) Treatment of the "singular situation": $T'_x(x,s)$ singular, or nearly singular, for one point $(x(s_n),s_n)$ of the homotopy path. There are some ideas on how to avoid those points in [3] , [6] or use methods of bifurcation theory ([5] , [8] , [9] , [17]).

(3) Selection of an optimal stepsize assuring attainment of the solution $x=x(1)$ within a required accuracy and with minimal effort. This paper will deal only with this last type of problem.

First results were obtained with the aid of theorems giving the feasibility of the above mentioned step-methods under certain (sufficient) conditions, e.g. 1 , 17 for the

Newton method or [19] for the method of the minimal residual. Compare also theorem 1 and 3 of this paper. The bounds S_o for Δs computed there are usually maximum bounds in that sense that $x(s_n)$ - more precisely, the last approximation $x_n^{k_n}$ for $x(s_n)$ - is just a point of attraction of the neighbouring point $x(s_{n+1})$ with $s_{n+1}-s_n < S_o$ for $n=o,1,\dots,N-1$. S_o is dependent on global constants such as $\|T'_x(x,s)\|$, $\|T'_s(x,s)\|$ $\|T''_x(x,s)\|$, $\|T'_x(x,s)^{-1}\|$.

Rheinboldt |14| suggests estimating the local convergence radius at each level s_n thus obtaining an estimate of the local (maximum) stepsize $\Delta s_n = s_{n+1}-s_n$.

Leder [11] extends the homotopy equation to

$$H(z) := H(x,s) := \begin{Bmatrix} T(x,s) \\ m(1-s) \end{Bmatrix} = \begin{Bmatrix} o \\ o \end{Bmatrix} \quad (m > o \text{ fixed})$$

and minimizes $\frac{1}{2}\|\phi(u)\|_2^2$ (using for instance, the method of Goldstine-Armijo). The stepsize control is done there locally and automatically.

Deuflhard [4] estimates, using an idea similar to that of Rheinboldt, the maximal possible stepsize Δs_n with the aid of local constants. The starting point for the (M.N.V.) is obtained by computing a predictor step of order p. He uses the damped Newton method in a corrector step. Then one can obtain an estimate for $(\Delta s_n)_{max}$. If necessary this stepsize is decreased by a certain factor.

All those ideas have in common the aim of obtaining a stepsize Δs_n as large as possible. (This is also true for Leder's method, where the maximization of Δs_n is combined with the minimization of the residual).This philosophy of employing the maximum stepsize is problematic. It has its foundation in the hypothesis that the total computaional effort is proportional to the number N of steps. In the above papers the effort necessary for the computation of $x(s_{n+1})$ with starting vector $x(s_n)$ by means of the local method $(x(s_n) \to x(s_{n+1}),\ k_{n+1}$ iterations) was not regarded.

Contrary to this in our paper we assume the computational effort to be approximately proportional to $k_1+k_2+\ldots+k_N$. The number k_n of the local iterations depends on the quality of the approximation $x_n^{k_n}$ of $x(s_{n+1})$, that means: large steps will cause large k_n and, conversely, small steps will allow small k_n. The minimum need not be at $\Delta s \approx S_o$ (where S_o denotes the possible maximum stepsize). These considerations were carried out for some iteration methods of the type of minimal residual (Wacker [19]) and will now be applied to (N.V.) and (M.N.V.).

Theorem 1 gives the theoretical globalization of (N.V.). In Theorem 2 we propose a version which can be implemented and for which the computational effort is minimized. More explicitly: the effort of constructing a possible initial value $x_{N-1}^{k_{N-1}}$ for the original problem $T(x,1)=0$ is minimized. Theorem 3 and 4 give the corresponding results for (M.N.V.). The formulation allows a solution independent of the specific problem and gives "natural constants" of (N.V.) and (M.N.V.). In the case that $S_o > 1$, we could interpret the results as "acceleration of convergence with the help of imbedding". This is in contrast to the common belief that imbedding should be used only if an initial guess for the iteration method is not known. The numerical tests seem to confirm our model - even relative to the aspect of acceleration of convergence by imbedding.

One disadvantage of this point of view is the requirement that the values $\|T'_x(x,s)^{-1}\|$, $\|T''_x(x,s)\|$, $\|T'_s(x,s)\|$ in a certain domain be known (at least in a neighbourhood of the homotopy path). One can try to improve the results by using the knowledge of the local behaviour at (x_n^k, s_n). To be able to formulate the result mathematically the knowledge at least of $T''_x(x,s)$ in this neighbourhood seems to be necessary.
It would be desirable to obtain results concerning the exact shape of the "effort function" $\phi(\Delta s,\varepsilon)$ as a function of s with the help of local estimations for $\|T'_x(x(s_n),s_n)^{-1}\|$, $T''_x(x(s_n),s_n)\|$, $\|T'_s(x(s_n),s_n)\|$, $\|T'_x(x(s_n),s_n)\|$.

II. MINIMIZING THE COMPUTATIONAL EFFORT OF THE GLOBALIZED NEWTON METHOD

Let $T(z):=T(x,s)$ and $K(x_o,R):=\{x/\|x-x_o\|\leq R\}$.

Theorem 1:

Let for the operator T(z) the following assumptions be valid in $K(x_o,R)\times[o,1] =: K\times I$.

(1) $T(x_o,o) = o$

(2) $\bigvee_{A,B>o} \; \bigwedge_{z\in K\times I} \quad \|T'_s(z)\| \leq A \wedge \|T''_x(z)\| \leq B$

(3) $\bigvee_{M>o} \bigwedge_{z \in K\times I} \| T'_x(z)^{-1} \| \le M$

(4) for $\Delta s < S_o := \frac{2}{M^2BA}$, $R > AM(1+\Delta s \sum_{k=o}^{\infty} (\frac{\Delta s}{S_o})^{2^k-1})$

Then we have:

for the uniform stepsize Δs the local transition $x(s_n) \to x(s_{n+1})$ $(s_{n+1} := s_n + \Delta s)$ using the initial value $x^o_{n+1} := x(s_n)$ is possible with (N.V.) for all n.

Proof:

Local step from level n to level n+1: we have

$x^o_{n+1} := x(s_n) \in K(x_o, R_n)$ with $R_n := s_n AM$

(resulting from the Davidenko-differential equation:

$x'(s) = -T'_x(x(s),s)^{-1} \circ T'_s(x(s),s) \wedge x(o) = x_o$

$\Longrightarrow \| x(s_n) - x_o \| = \| \int_o^{s_n} T'_x(x(s),s)^{-1} \circ T'_s(x(s),s)ds \| \le AMs_n$)

A sufficient condition for the convergence of (N.V.) for the problem $T(x,s_{n+1}) = o$, $x^o_{n+1} := x(s_n)$ is (compare [1o] , page 579): $h := M^2B\eta < 2$ where $\| T(x^o_{n+1}, s_{n+1}) \| \le \eta$ and M,B are bounds for $\| T'_x(x,s_{n+1})^{-1} \|$, $\| T''_x(x,s_{n+1}) \|$ in a ball $K(x^o_{n+1}, r)$ with $r = M\eta \sum_{k=o}^{\infty} (\frac{h}{2})^{2^k-1}$.

$R > R_n + r$ ($\Longleftrightarrow K(x_o,R) \supset K(x^o_{n+1}, r)$) implies:

1) for M and B we can use the corresponding constants of assumption (2),resp. (3).

2) $\| T(x^o_{n+1}, s_{n+1}) \| = \| T(x(s_n), s_{n+1}) - T(x(s_n), s_n) \| \le A.\Delta s$.

   Therefore the requirement $h < 2$ leads to

   $M^2BA\Delta s < 2 \Longrightarrow \Delta s < S_o := \frac{2}{M^2BA}$.

That means: sufficient for the possibility of the local transition are the requirements:

$\Delta s < S_o$ and $R > R_n + r$, $r = o,1,\ldots,N-1$.

$R > R_n + r$, $n = o, 1, \ldots, N-1$ is valid if:

$$R > AM + r(\Delta s) \quad \text{with} \quad r(\Delta s) := r = MA\Delta s \sum_{k=o}^{\infty} \left(\frac{\Delta s}{S_o}\right)^{2^k - 1}$$

$$\Longleftrightarrow \quad R > AM\left(1 + \Delta s \sum_{k=o}^{\infty} \left(\frac{\Delta s}{S_o}\right)^{2^k - 1}\right)$$

□

Remark 1:

$$\sum_{k=o}^{\infty} \left(\frac{\Delta s}{S_o}\right)^{2^k - 1} \leq \sum_{k=o}^{\infty} \left(\frac{\Delta s}{S_o}\right)^{k} = \frac{1}{1 - \frac{\Delta s}{S_o}} = \frac{S_o}{S_o - \Delta s}$$

Therefore in Theorem 1 the condition $R > AM\left(1 + \frac{\Delta s S_o}{S_o - \Delta s}\right)$ in (4) is sufficient.

Remark 2:

We can modify the proof (varying slightly the condition on R) for arbitrary stepsizes $\Delta s_n \in (o, S_o)$, $\Delta s_n := s_{n+1} - s_n$ with

$$R > AM\left(1 + \Delta s \sum_{k=o}^{\infty} \left(\frac{\Delta s}{S_o}\right)^{2^k - 1}\right), \quad \Delta s := \max_{o \leq n \leq N-1} \Delta s_n$$

In Theorem 1 we assumed that the intermediate solutions $x(s_n)$ are computed exactly, that means we assumed them to lie on the homotopy path $x(s)$. This will not be possible in actual computation. The following implementable version seems reasonable:

Given a number of pairs $(s_o, \varepsilon_o), (s_1, \varepsilon_1), \ldots, (s_N, \varepsilon_N)$ with $o = s_o < \ldots < s_N = 1$ and $\varepsilon_i > o$.

At level n we iterate just k_n times such that for the last iteration $x_n^{k_n}$ of this level the following is valid:

$$r_n := \| T(x_n^{k_n}, s_n) \| \leq \varepsilon_n$$

The initial vector for the next level n+1: $x_{n+1}^{o} := x_n^{k_n}$

We start the process with $x_1^o := x_o$ as initial value of the first level with $r_o := \| T(x_o, o) \| \leq \varepsilon_o$

Aim: Given a final accuracy ε_N: A finite sequence $\{s_i\}$ partitioning the homotopy interval $[o,1]$ as well as the corresponding accuracies $\{\varepsilon_i\}$ $(i \le N-1)$ are to be found such that the algorithm works and that the complete computational effort to obtain the solution $\hat{x} := x(1)$ with an accuracy $\|T(\hat{x},1)\| \le \varepsilon_N$ is minimal.

We simplify the problem in the following way:

1) To minimize the computational effort we only regard the total numbers of the iterations of all levels $k_1+\ldots+k_N$
2) We assume to know global bounds for $\|T''_x(z)\|$, $\|T'_x(z)^{-1}\|$, $\|T'_s(z)\|$.
3) We only use uniform stepsizes Δs (i.e. $s_n := n\Delta s$).

A partial solution of this problem is Theorem 2

Theorem 2:

For the operator $T(z)$ in $K(x_o,R) \times [o,1] =: K \times I$ the following assumptions be valid:

(1) $\|T(x_o,O)\| \le \varepsilon$

(2) $\bigvee_{A,B>o} \quad \bigwedge_{z \in K \times I} \quad \|T'_s(z)\| \le A \wedge \|T''_x(z)\| \le B$

(3) $\bigvee_{M>o} \quad \bigvee_{z \in K \times I} \quad \|T'_x(z)^{-1}\| \le M$

(4) $\varepsilon < S_o A \; ; \; S_o := \dfrac{2}{M^2 BA}$

(5) for $\Delta s < S_o - \frac{\varepsilon}{A}$, $R > AM(1+(\Delta s + \frac{\varepsilon}{A}) \sum_{k=o}^{\infty} (\frac{\Delta s+\frac{\varepsilon}{A}}{S_o})^{2^k-1})$

Then we have

1) The globalized Newton method in the implementable version works with the uniform stepsize Δs and $\varepsilon_n := \varepsilon$, $n=o,1,..,N-1$ and with a given final accuracy $\varepsilon_N > o$.

2) The following recursion is valid for the residua r_n:

$$r_{n+1} \le \left(\frac{BM^2}{2}\right)^{-1} \left[\left(\frac{BM^2}{2}\right)(A\Delta s + r_n)\right]^{2^{k_{n+1}}}$$

3) To assure $r_n \le \varepsilon\,(n \le N-1)$, $\bar{k} := [k(\Delta s,\varepsilon)]$ iterations at each level are sufficient, where

$$k(\Delta s,\varepsilon) := \ln\left\{\frac{2.\ln(\varepsilon \frac{M^2 B}{2})}{\ln((\frac{M^2 B}{2})(A\Delta s+\varepsilon))}\right\}$$

4) The quantity $\phi(\Delta s,\varepsilon) := \frac{1}{\Delta s}\,k(\Delta s,\varepsilon)$ is an upper bound for the total number of iterations from level 1 to level N-1. At $(\Delta\bar{s},\bar{\varepsilon}) := (S_o\bar{u}, S_o A\bar{v})$ the upper bound for the effort $\phi(\Delta s,\varepsilon)$ assumes its absolute minimum. $(\bar{u},\bar{v})$ is the (unique) solution of the system

$$v.\ln(v)(\ln(2)(u+v)-\ln(v))+u = o$$

$$\ln(u+v)(u+v) = \ln(v).v$$

(numerically $\bar{u} \approx o.619$, $\bar{v} \approx o.1o4$ and $\phi(\Delta\bar{s},\bar{\varepsilon}) \approx \frac{1}{S_o}\,4.258$)

Proof:

1) Implementation of the total process.

a) starting the method $(s=s_1)$ the initial vector x_1^o equals x_o and we know by assumption (1): $x_1^o \in K(x_o,R_o)$ with $R_o = o \wedge r_o := \|T(x_o,o)\| \le \varepsilon$

b) generally we see:

Let $x_{n+1}^o \in K(x_o,R_n) \wedge R \ge R_n \wedge r_n \le \varepsilon$. Then we have:

$$\|x_{n+1}^o - x(s_n)\| \le M\,\|T(x_{n+1}^o,s_n) - T(x(s_n),s_n)\| = M.r_n \le M.\varepsilon$$

$$\Longrightarrow \|x_{n+1}^o - x_o\| \le \|x_{n+1}^o - x(s_n)\| + \|x(s_n)-x_o\| \le MAs_n + M\varepsilon =: \bar{R}_n$$

$$\Longrightarrow x_{n+1}^o \in K(x_o,\bar{R}_n)$$

and further:

$$\|T(x^o_{n+1},s_{n+1})\| \le \|T(x^o_{n+1},s_{n+1})-T(x^o_{n+1},s_n)\| + $$
$$+ \|T(x^o_{n+1},s_n)\| \le A.\Delta s+\varepsilon$$

Analogous to Theorem 1 we have:

If $R > \bar{R}_{n+1}$ ($\Longrightarrow$ $K(x_o,R) \supset K(x^o_{n+1},r)$), the Newton method for $T(x,s_{n+1})=o$ with initial value $x^o_{n+1}:=x_n^{k_n}$ $(r_n \le \varepsilon)$ converges, if $h < 2 \iff M^2B(A\Delta s+\varepsilon) < 2 \iff \Delta s < S_o - \frac{\varepsilon}{A}$ and the iterations x^k_{n+1} are in $K(x^o_{n+1},r) \subset K(x_o,R_{n+1})$ with

$$R_{n+1}=\bar{R}_n+r, \quad r=M\eta \sum_{k=o}^{\infty} (\frac{h}{2})^{2^k-1} = M(A\Delta s+\varepsilon) \sum_{k=o}^{\infty} (\frac{M^2B(A\Delta s+\varepsilon)}{2})^{2^k-1}$$

This shows that after a finite number k_{n+1} of iterations one has again: $r_{n+1} \le \varepsilon$.

To guarantee the feasibility of continuation the conditions $\Delta s < S_o - \frac{\varepsilon}{A}$ and $R > \bar{R}_n+r$, $n=o,1,\dots,N-1$ are sufficient.

The second condition is valid if:

$$R > AM+r = AM+M(A\Delta s+\varepsilon) \sum_{k=o}^{\infty} (\frac{M^2B(A\Delta s+\varepsilon)}{2})^{2^k-1} =$$

$$= AM(1+(\Delta s+\frac{\varepsilon}{A}) \sum_{k=o}^{\infty} (\frac{\Delta s-\frac{\varepsilon}{A}}{S_o})^{2^k-1})$$

2) Recursive law for the residua.

For the residua of level n+1 we have (see |1o| , p.579):

$$r_{n+1} = \|T(x^{k_{n+1}}_{n+1},s_{n+1})\| \le (\frac{BM^2}{2})^{-1} \left[\frac{BM^2}{2}\|T(x^o_{n+1},s_{n+1})\|\right]^{2^{k_{n+1}}}$$

$$\|T(x^o_{n+1},s_{n+1})\| = \|T(x_n^{k_n},s_{n+1})-T(x_n^{k_n},s_n)\| + \|T(x_n^{k_n},s_n)\| \le$$
$$\le A\Delta s+r_n$$

and therefore:

$$r_{n+1} \le (\frac{BM^2}{2})^{-1} \left[\frac{BM^2}{2}(A\Delta s+r_n)\right]^{2^{k_{n+1}}}$$

3) Computation of $\bar{k}$:

Performing at least $\bar{k} = [k(\Delta s,\varepsilon)]$ iterations at each level we obtain:

$r_{n+1} \leq \phi(r_n)$ with $\phi(x) := (\frac{BM^2}{2})^{-1} (\frac{BM^2}{2} (A\Delta s+x))^{2^{\bar{k}}}$

(compare (2))

Then we have

$r_o \leq \varepsilon \implies r_1 \leq \phi(r_o) \leq \phi(\varepsilon) \leq \varepsilon$

ϕ monotone (first $\leq$); the last $\leq$ follows from the choice of $\bar{k}=[k(\Delta s,\varepsilon)]$.

$\implies r_1 \leq \varepsilon$

Analogously we get $r_2 \leq \varepsilon, \ldots$ and finally $r_{N-1} \leq \varepsilon$.

4) a) Computing an upper bound for the computational effort:

Because of 3) we get that $(N-1)\bar{k}$ is an upper bound for the iterations necessary to get from level 1 to level N-1.

$$(N-1)\bar{k} \leq \frac{1}{\Delta s} k(\Delta s,\varepsilon) =: \Phi(\Delta s,\varepsilon)$$

we restrict ourselves to the discussion of this upper bound $\Phi(\Delta s,\varepsilon)$.

b) Minimization of the upper bound $\Phi(\Delta s,\varepsilon)$:

The continuation in the implementable version can be performed if

$$M^2B(A\Delta s+\varepsilon)<2 \iff (\Delta s,\varepsilon)\in D:=\{(\Delta s,\varepsilon)/\frac{M^2B}{2}(A\Delta s+\varepsilon)<1 \wedge \Delta s>o \wedge \varepsilon>o\}$$

The minimization problem can be presented in the following form which is (almost) independent of the respective concrete problem:

$$u:=\frac{M^2BA}{2}\Delta s \ , \ v:=\frac{M^2B}{2}\varepsilon$$

$$\Phi(\Delta s,\varepsilon) = \min! \wedge (\Delta s,\varepsilon)\in D \iff \Phi_1(u,v) = \min! \wedge (u,v)\in D_1$$

with $\Phi_1(u,v) = \frac{1}{u}\ln\left\{\frac{2\ln(v)}{\ln(u+v)}\right\} \wedge D_1:=\{(u,v)/u+v<1 \wedge u>o \wedge v>o\}$

On ∂D_1 the following property is valid: $\bigwedge_{(u,v)\in\partial D_1} \Phi_1(u,v)=+\infty$

and in the interior: $\bigwedge_{(u,v)\in \mathring{D}_1} \Phi_1(u,v) \geq 0$

$\Longrightarrow$ $\Phi_1(u,v)$ has a (unique) minimum in $\mathring{D}_1$:

Necessary conditions for this minimum are:

$$\frac{\partial\Phi_1}{\partial u} = \left\{\frac{-u}{(u+v)\ln(u+v)} - \ln\left[\frac{2\ln(v)}{\ln(u+v)}\right]\right\} /u^2 = 0$$

$$\frac{\partial\Phi_1}{\partial v} = \left\{\frac{(u+v)\ln(u+v)-v\ln(v)}{v\ln(v)(u+v)\ln(u+v)}\right\}/u = 0$$

Numerical solution: $z:=u+v$ leads to the system

$$-z+\ln(v)-v\ln(v)[\ln(2z)-\ln(v)] = 0$$

$$z\ln(z) - v\ln(v) = 0$$

The solution was computed numerically (by imbedding), where the the residual was less than 10^{-15}): $\begin{pmatrix}\bar{u}\\ \bar{v}\end{pmatrix} \approx \begin{pmatrix}0.618955137\\ 0.103558206\end{pmatrix}$

Thus we have

$$\begin{pmatrix}\Delta\bar{s}\\ \bar{\varepsilon}\end{pmatrix} = \begin{pmatrix}S_0\bar{u}\\ AS_0\bar{v}\end{pmatrix} \quad \text{and} \quad \Phi(\Delta\bar{s},\bar{\varepsilon}) = \frac{1}{S_0}\Phi_1(\Delta\bar{s},\bar{\varepsilon}) \approx \frac{1}{S_0} 4.258384683$$

□

Remark 3:

Since $\sum_{k=0}^{\infty} \left(\frac{\Delta s+\frac{\varepsilon}{A}}{S_0}\right)^{2^k-1} \leq \sum_{k=0}^{\infty} \left(\frac{\Delta s+\frac{\varepsilon}{A}}{S_0}\right)^{k} = \frac{S_0}{S_0-\Delta s-\frac{\varepsilon}{A}}$ in Theorem 2

the assumption

$R > AM\left(1+\frac{S_0(\Delta s+\frac{\varepsilon}{A})}{S_0-\Delta s-\frac{\varepsilon}{A}}\right)$ in (5) is sufficient.

Remark 4:

1) Theorem 2 gives us the possibility to minimize the computational effort for all iterations from level 1 to level N-1 (to be exact: we minimize an upper bound of this effort) by a suitable choice of $\Delta\bar{s} := S_0\bar{u}$ and $\bar{\varepsilon} := S_0A\bar{v}$.

The minimization is performed only for those globalizations using constant stepsize Δs and constant accuracy $\varepsilon_n = \varepsilon$ at all intermediate levels $(n=0,1,\ldots,N-1)$.

2) More precisely: not the effort for the whole process is minimized but the effort for finding a feasible initial value for the last step, that means for the original problem $T(x,1) \equiv T(x) = 0$.

III. MINIMIZING THE COMPUTATIONAL EFFORT FOR THE GLOBALIZED MODIFIED NEWTON METHOD

Theorem 3:

Let for $T(z)$ the following assumptions be valid in $K(x_o,R) \times I =: K \times I$.

(1) $T(x_o,o) = o$

(2) $\bigvee_{A,B>0} \ \bigvee_{z \in K \times I} \ \|T'_s(z)\| \leq A \wedge \|T''_x(z)\| \leq B$

(3) $\bigvee_{M>o} \ \bigwedge_{z \in K \times I} \ \|T'_x(z)^{-1}\| \leq M$

(4) for $\Delta s < S_o := \dfrac{1}{2M^2BA}$, $R > AM(1+2S_o(1-\sqrt{1-\frac{\Delta s}{S_o}}))$

Then we have:

For the uniform stepsize Δs the local continuation $x(s_n) \to x(s_{n+1})$, $s_{n+1}:=s_n+\Delta s$, can be performed using (M.N.V.) with initial value $x^o_{n+1}:=x(s_n)$.

Sketch of the proof:

(analogous to Theorem 1; but here we refer to the local version [1o], p.572).

Local continuation from level n to n+1:

A sufficient condition for the convergence of (M.N.V.) for $T(x,s_{n+1})=o$ with $x^o_{n+1}:=x(s_n)$ is $h:=M^2B\eta < \frac{1}{2}$ that leads to $\Delta s < S_o := 1/2M^2BA$. To guarantee the performance for all local transitions it is sufficient to require $\Delta s < S_o$ and $R > R_n+r$, $n=o,1,\dots,N-1$ with

$$r = r(\Delta s) = \left[1-\sqrt{1-2M^2BA\Delta s}\,\right]/MB$$

$$R > R_n+r \text{ valid if } R > AM+r(\Delta s) = \left(BAM^2+\left[1-\sqrt{1-2M^2BA\Delta s}\right]\right)/MB$$

□

Remark 5:

Because of $AM+r(\Delta s) < AM+r(S_o) = AM+1/MB$ in Theorem 3 the property $R > AM+1/MB = AM(1+2S_o)$ is sufficient in (4).

Remark 6:

Just as for Theorem 1 it is possible here to reformulate the above Theorem for arbitrary stepsizes $\Delta s_n \in (o,S_o)$, $\Delta s_n:=s_{n+1}-s_n$ with $R > AM+r(\Delta s)$, $\Delta s:= \max_{o\le n\le N-1} \Delta s_n$.

In the same way as in Theorem 2 we want to minimize the computational effort for an implementable version of Theorem 3. That means: for a given final accuracy ε_N we look for a uniform partition of the homotopy interval and a corresponding accuracy ε (level o to level N-1) such that

a) the algorithm works

b) the total computational effort for the computation of an initial value x_N^o for the original problem will be minimized.

The global quantities $\|T_x''(z)\|$, $\|T_x'(z)^{-1}\|$, $\|T_s'(z)\|$ are assumed to be known: We only regard the total number of iterations from level 1 to level N-1.

Lemma:

(estimate for the residual in M.N.V.)

Let for the operator T(x) the following assumptions be valid in $K(x_o,R)=:K$

(1) $\|T(x_o)\| \leq \eta \wedge \|T'(x_o)^{-1}\| \leq M$

(2) $\bigwedge_{x \in K} \|T_x''(x)\| \leq B$

(3) $h:=M^2B\eta < \frac{1}{2}$

Then we have for M.N.V. $(x_{n+1} = x_n - T'(x_o)^{-1}(x_n))$:

$\|T(x_n)\| \leq \eta(1-\sqrt{1-2h})^n$

Proof:

We regard (M.N.V.) to be a successive approximation for the fixed point equation $x=x-T'(x_o)^{-1}T(x) =: S(x)$.

Let $\phi(t) := \frac{1}{2}MBt^2+M\eta, t\ [t_o,t_o+r] := [o,R]$

$\phi(t)$ fulfills the assumptions of Theorem 1 [1o] , p 564.

Therefore we have $\|x_n-x_{n-1}\| \leq t_n-t_{n-1}$, $t_n:=\phi(t_{n-1})$

(compare proof of Theorem 1 [1o] , p.564).

$$\|T(x_n)\| = \|T'(x_o)(x_{n+1}-x_n)\| = \|\int_{x_{n-1}}^{x_n} T'(x_o)S'(x)dx\|$$

$$\|T'(x_o)S'(x)\| = \|T'(x_o)-T'(x)\| \leq B\|x-x_o\| \leq Bt = \frac{1}{M}\phi'(t)$$

for $t-t_o \geq \|x-x_o\|$

Then we use a result in [1o] , p.541 :

$$\|T(x_n)\| = \|\int_{x_{n-1}}^{x_n} T'(x_o)S'(x)dx\| \le \frac{1}{M}\int_{t_{n-1}}^{t_n} \phi'(t)dt = \frac{1}{M}(\phi(t_n) -$$

$$-\phi(t_{n-1})) = \frac{1}{M}(t_{n+1}-t_n) \quad \text{because of} \quad \|x_n-x_{n-1}\| \le t_n-t_{n-1}$$

$$\Longrightarrow M\|T(x_n)\| \le t_{n+1}-t_n$$

The smallest fixed point of $\phi(t)$ is $\bar{t} := \dfrac{1-\sqrt{1-2h}}{MB}$ $\quad (h:=M^2B\eta)$

We have $t_n \le \bar{t}$ (compare Theorem 1, [1o] p.564)

Then we have:

$$t_{n+1}-t_n = \phi(t_n)-\phi(t_{n-1}) = \tfrac{1}{2}MB(t_n+t_{n-1})(t_n-t_{n-1}) \underset{t_k \le \bar{t}}{\le} MB\bar{t}(t_n-t_{n-1}) =$$

$$= (1-\sqrt{1-2h})(t_n-t_{n-1})$$

therefore

$$M\|T(x_n)\| \le (1-\sqrt{1-2h})^n(t_1-t_o) = M\eta(1-\sqrt{1-2h})^n$$

$$\Longrightarrow \|T(x_n)\| \le \eta(1-\sqrt{1-2h})^n$$

□

Theorem 4:

Let for $T(z)$ the following assumptions be valid in $K(x_o,R) \times [o,1] =: K \times I$

(1) $\|T(x_o,o)\| \le \varepsilon_o$

(2) $\bigvee_{A,B>o} \; \bigwedge_{z\in K\times I} \; \|T_s'(z)\| \le A \wedge \|T_x''(z)\| \le B$

(3) $\bigvee_{M>o} \; \bigwedge_{z\in K\times I} \; \|T_x'(z)^{-1}\| \le M$

(4) $\varepsilon < AS_o$, $S_o := \dfrac{1}{2M^2BA}$

(5) for $\Delta s < S_o - \frac{\varepsilon}{A}$, $R > AM(1+2S_o(1-\sqrt{1-\frac{\Delta s+\varepsilon/A}{S_o}}) + \frac{\varepsilon}{A})$

Then we have:

1) The implementable version of the globalized modified Newton method works with the uniform stepsize Δs and $\varepsilon_n = \varepsilon$ for $n=o,1,\ldots,N-1$ and with a given final accuracy $\varepsilon_N > o$.

2) The following recursion is valid for the residua r_n:

$$r_{n+1} \leq (1-\sqrt{1-2M^2B(A\Delta s+r_n)})^{k_{n+1}} (A\Delta s+r_n)$$

3) To guarantee $r_n \leq \varepsilon$ $(n \leq N-1)$, $\bar{k} := k(\Delta s,\varepsilon)$ iterations at each level are sufficient

$$k(\Delta s,\varepsilon) := \frac{\ln(\frac{\varepsilon}{A\Delta s+\varepsilon})}{\ln(1-\sqrt{1-2M^2B(A\Delta s+\varepsilon)})} + 1$$

4) The total number of the iterations from level 1 to level N-1 is less (or at most equal) than $\Phi(\Delta s,\varepsilon) := \frac{1}{\Delta s} k(\Delta s,\varepsilon)$. At $(\Delta\bar{s},\bar{\varepsilon}) := (S_o\bar{u}, S_oA\bar{v})$ the upper bound for the computational effort assumes its absolute minimum. $(\bar{u},\bar{v})$ denotes here the solution of the system

$$u2w(1-w)\ln(1-w)+(u+v)\ln(\frac{1}{u+v})+\ln(1-w)\ln(\frac{v}{u+v})+\ln(1-w)2w(1-w)(u+v)=o$$

$$u2w(1-w)\ln(1-w)-v(u+v)\ln(\frac{v}{u+v}) = o$$

where $w:=\sqrt{1-u-v}$

(numerical solution $\bar{u} \approx o.452$, $\bar{v} \approx o.245$ and $\Phi(\Delta\bar{s},\bar{\varepsilon}) \approx \frac{1}{S_o}$ 5.1o5

Proof:

(in the essential parts analogous to Theorem 2)

1) Implementation of the total process

Using Theorem 6 ([1o] , p.527) we obtain the following conditions to make the implementation of the process possible :

$2M^2B(A\Delta s+\varepsilon) < 1$ and $R > \bar{R}_n+r$, $r=o,1,\ldots,N-1$

$2M^2B(A\Delta s+\varepsilon) < 1 \iff \Delta s < S_o - \frac{\varepsilon}{A} \iff (\Delta s,\varepsilon) \varepsilon D$ with

$D = \{(\Delta s,\varepsilon) / 2M^2B(A\Delta s+\varepsilon) < 1 \wedge \Delta s > o \wedge \varepsilon > o\}$

$R > \bar{R}_n+r$ is valid if:

$$R > AM+M\varepsilon+r(\Delta s) = AM+M\varepsilon+(1-\sqrt{1-2M^2B(A\Delta s+\varepsilon)})/MB =$$

$$= AM(1+\frac{\varepsilon}{A} + 2S_o(1-\sqrt{1-\frac{\Delta s+\varepsilon/A}{S_o}}))$$

2) Recursive law for the residua:

Because of the assumptions of the Theorem we can apply the Lemma and get

$$\| T(x_{n+1}) \| = r_{n+1} \leq (1-\sqrt{1-2M^2B(A\Delta s+r_n)})^{k_{n+1}} (A\Delta s+r_n)$$

3) Computation of $\bar{k}$:

Having at least $\bar{k} = [k(\Delta s,\varepsilon)]$ iterations at each level, we get $r_{n+1} \leq \phi(r_n)$ with $\phi(x) := [1-\sqrt{1-2M^2B(A\Delta s+x)}]^{\bar{k}}(A\Delta s+x)$

This implies (with the same arguments as in Theorem 2)

$r_n \leq \varepsilon$, n=o,1,...,N-1.

4) Computing an upper bound for the computational effort:

a) we use the following upper bound:

$$\phi(\Delta s,\varepsilon) := \frac{k(\Delta s,\varepsilon)}{\Delta s} = \left\{ \frac{\ln(\varepsilon/(A\Delta s+\varepsilon)}{\ln(1-\sqrt{1-2M^2B(A\Delta s+\varepsilon)})} +1 \right\} / \Delta s$$

like in the proof of Theorem 2 we are able to give a formulation which is (almost) independent of the respective concrete situtation:

$u := 2M^2BA\Delta s, \; v := 2M^2B\varepsilon$

$$\phi(\Delta s,\varepsilon) = \min! \wedge (\Delta s,\varepsilon) \in D \iff \phi_1(u,v) = \min! \wedge (u,v) \in D_1$$

with $$\phi_1(u,v) := \left\{ \frac{\ln(\frac{v}{u+v})}{\ln(1-\sqrt{1-u-v})} + 1 \right\} /u$$

$D_1 := \{(u,v) / u+v < 1 \wedge u > o \wedge v > o\}$

b) Minimization of the upper bound:

$$\bigwedge_{(u,v)\in \partial D_1} \Phi_1(u,v) = +\infty \quad \text{and} \quad \bigwedge_{(u,v)\in \mathring{D}_1} \Phi_1(u,v) \geq o$$

$\Longrightarrow$ $\Phi_1(u,v)$ has a (unique) minimum in D_1.

Necessary conditions for this minimum are $(w:=\sqrt{1-u-v})$

$$\frac{\partial\Phi_1}{\partial u} = \left\{ \frac{u - \left[\frac{\ln(1-w)}{(u+v)} - \frac{\ln(\frac{v}{u+v})}{2w(1-w)}\right]}{\ln^2(1-w)} - \frac{\left[\ln(\frac{v}{u+v}) + \ln(1-w)\right]}{\ln(1-w)} \right\} \Big/ u^2 = 0$$

$$\frac{\partial\Phi_1}{\partial v} = \left\{ \frac{\frac{\ln(1-w)u}{v(u+v)} - \ln(\frac{v}{u+v}) \cdot \frac{1}{2w(1-w)}}{u \ln^2(1-w)} \right\} = 0$$

This leads to the system:

$$u\left[2w(1-w)\ln(1-w) + (u+v)\ln(\tfrac{v}{u+v})\right] + \ln(1-w)\left[\ln(\tfrac{v}{u+v}) + \ln(1-w)\right]2w(1-w)(u+v) = 0$$

$$u2w(1-w)\ln(1-w) - v(u+v)\ln(\tfrac{v}{u+v}) = 0$$

or after some calculations

$$(u+v)^2\ln(\tfrac{v}{u+v}) + \ln(1-w)\left[\ln(\tfrac{v}{u+v}) + \ln(1-w)\right]2w(1-w)(u+v) = 0$$

$$u2w(1-w)\ln(1-w) - v(u+v)\ln(\tfrac{v}{u+v}) = 0$$

Solution: $\bar{u} \approx 0.45218749$, $\bar{v} \approx 0.24523429$

we get $\begin{pmatrix} \Delta\bar{s} \\ \bar{\varepsilon} \end{pmatrix} = \begin{pmatrix} S_0\bar{u} \\ AS_0\bar{v} \end{pmatrix}$

The complete number of iterations is less (or equal)

$$\Phi = \frac{1}{S_0}\,\Phi_1(\bar{u},\bar{v}) \approx \frac{1}{S_0} \cdot 5.10552170 5$$

Remark 7:

Because of $\Delta s < S_0 - \frac{\varepsilon}{A}$ in Theorem 4 the assumption $R > AM(1 + \frac{\varepsilon}{A} + 2S_0)$ in (5) is sufficient.

Remark 8:

Theorem 2 respectively 4 tells us that at most $\bar{k}$ iterations per level are enough to guarantee $\|T(x_n^{\bar{k}}, s_n)\| \le \bar{\varepsilon}\,(n \le N-1)$. Actually in the practical computation we pass to the next level at $k_n \le \bar{k}$ if $r_n \le \bar{\varepsilon}$ is valid. With exception of the last level we use the same accuracy $\bar{\varepsilon}$ at each level. Compare in this context the method suggested by Schwetlick [15],p.216 ff. There the stepsize is controlled without global constants only by checking the residua: $\|T(x^k_{n+1},s_{n+1})\| \le \|T(x^k_n,s_n)\|+\varepsilon\Delta s_n$

In the paper of Schwetlick the accuracy for the first part of the sequence (i.e. $x_o,x_1,..,x_{N-1}$) is higher (at least theoretically) then here and therefore too much work has to be done during this first part. (In the same paper an interesting modification is proposed, which works with a family of imbeddings using the result of one imbedding as the starting value of the following one). See figure below.

Remark 9:

a) The constants $(\bar{u},\bar{v})$ received from Theorems 2 and 4 are not depending on the problem and one might think of them

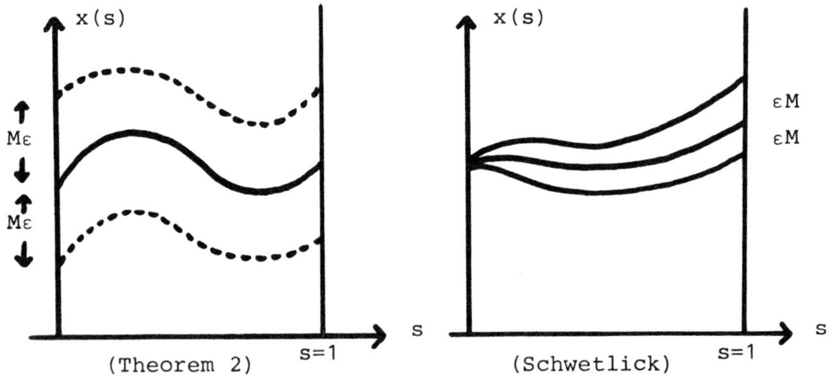

$\|\Delta x_n\| \cong M\|r_n\|$

as "natural constants" of the globalized version of (N.V.) respectively (M.N.V.).

b) Theorem 2 and 4 resp. shows an analysis of minimizing the computational effort for imbedding methods.
In spite of the severe simplifications one can expect that the obtained results namely $\Delta\bar{s} = \bar{u}s_o \approx \frac{s_o}{2}$, *(i.e. Δs not maximal in the feasible interval)* and the possibility to get influence by choosing a suitable intermediate accuracy $\bar{\varepsilon}$ *(on the one side even at a high final accuracy ε_N a poorer accuracy ε_n on each level n, n=o(1)(N-1), will do; on the other side one can save computational effort using a higher accuracy on each intermediate level, compare Theorem 2 resp. 4 in the case of $\bar{\varepsilon} < \varepsilon_N$)* can be realized at concrete numerical problems. See for this the numerical tests in chapter 4.
Difficulties arising there are: The global constants A,B, M are difficult to find and the estimates for Δs are usually very poor, if one works with them. To improve the results one could try to use local constants (A_n, B_n, M_n) but always keeping a global optimization in mind. Compare [4] and [13] where the control of the stepsize with the aid of local constants was suggested.

c) A natural generalization of Theorem 2 resp. 4 consists in the discussion of a variable stepsize Δs_n and accuracy ε_n at each level. This is possible with the same theoretical instruments used in this paper though one looses the explicite character of our solution. Using other methods we could get and solve a model [2o] which confirms the results of the model proposed here and which is further adapted to a numerical implementation.

IV. NUMERICAL TESTS

The following examples demonstrate qualitatively the statements of Theorem 2 and 4.

Example 1:

$$\begin{aligned} x_1 &= x_1 - \tfrac{1}{3} e^{-x_2} + \tfrac{1}{3e} \\ x_2 &= \tfrac{x_2}{2} + \tfrac{1}{5} \ln(1 + x_3^2) + \tfrac{1}{2} - \tfrac{1}{5} \ln(2) \qquad \Longleftrightarrow \quad x = A(x) \\ x_3 &= x_3 - \sin(x_1) + \sin(1) \end{aligned}$$

Linear imbedding: $x = sA(x)$. Solution: $\bar{x} = (1,1,1)^T$

With fixed $\varepsilon_N := 10^{-4}$ as final accuracy the problem was computed with different Δs and different ε(intermediate accuracy).

$It(\varepsilon)$: number of iterations at an intermediate accuracy ε

$k_1(k_2)$: number of iterations for step 1 until N-1 (step 1 until N)

Div(k): after k iterations at a fixed level the desired accuracy ε, ε_N resp., was not attained, the process stopped.

(N.V.):

Δs	0.1	0.2	0.3	0.4	0.5	0.6	0.7	0.8
It(10^{-6})	24(31)	11(16)	10(17)	7(11)	3(7)	4(7)	4(7)	5(9)
It(10^{-4})	20(27)	9(14)	8(15)	7(11)	3(7)	3(6)	4(7)	4(8)
It(10^{-2})	10(16)	6(10)	5(11)	5(9)	2(6)	2(5)	3(6)	3(7)

Δs	0.9	0.95	0.955	0.956	0.9562	0.95621	0.95622
It(10^{-6})	10(17)	19(25)	67(71)	76(50)	104(108)	131(135)	Div(500)
It(10^{-4})	9(16)	18(25)	67(71)	75(79)	103(107)	131(135)	Div(500)
It(10^{-2})	8(13)	17(25)	93(97)	Div(500)	——	——	——

(M.N.V.):

Δs	0.1	0.2	0.3	0.4	0.5	0.6	0.7	0.8	0.9	0.90072
It(10^{-6})	38(56)	18(49)	21(39)	17(35)	8(18)	11(20)	19(29)	14(32)	18(36)	Div(250)
It(10^{-4})	25(43)	12(31)	14(32)	12(30)	5(15)	8(17)	12(22)	10(28)	14(32)	——
It(10^{-2})	10(17)	6(14)	5(21)	5(26)	2(12)	4(14)	6(16)	6(25)	9(25)	——

Observations:

(1) It can be clearly seen that the minimum of the computational effort is not attained at the largest possible stepsize. The same behaviour was observed throughout all our numerical tests (only few of them are listed here). In contrary to Theorem 2 and 4 even several local minima are possible.

(2) For our example we get experimentically: $S_o \approx 1$.
The theory of our model yields $\Delta s \approx 0.5\ S_o$. The numerical tests give at least approximately this value too.

(3) The saving of the computational effort in some cases is considerable:

e.g. (N.V.) $It(10^{-4})$: $\Phi(\overline{\Delta s}=0.6) = 3(6)$

$\Phi(\overline{\Delta s}=0.95621) = 131(135)$

(M.N.V.) $It(10^{-4})$: $\Phi(\overline{\Delta s}=0.5) = 5(15)$

$\Phi(\overline{\Delta s}=0.9) = 14(32)$

Theoretically the optimal values are:

(N.V.) : $\bar{\Phi} \approx \frac{4.258}{S_o} \approx 4.5$

(M.N.V.) : $\bar{\Phi} \approx \frac{5.165}{S_o} \approx 5.5$

Acceleration of Convergence:

Until now continuation methods have been used only in case where no starting points were known. Theorems 2 and 4 suggest the possibility of acceleration of convergence by continuation methods.

In the case $\Delta s < 1 < S_o$ $(\Longleftrightarrow S_o \in (1,\frac{1}{\alpha})$, where $\alpha_{N.V.} \approx 0.619$ and $\alpha_{M.N.V.} \approx 0.452)$ the computational effort is reduced by imbedding with stepsize $\overline{\Delta s}$ compared to the effort of solving the problem without imbedding. More precisely: the upper bound of the number of iterations (see Theorem 2 and 4) is smaller.

We give an example to demonstrate this fact:

Example 2:

$$\left.\begin{aligned} x_1 &= ax_1 - \frac{1}{3}e^{-x_2} - a + \frac{1}{3e} + 1 \\ x_2 &= bx_2 + \frac{1}{5}\ln(1+x_3^2) - b - \frac{1}{5}\ln(2) + 1 \\ x_3 &= cx_3 - \sin(x_1) - c + \sin(1) + 1 \end{aligned}\right\} \equiv x=A(x)$$

Imbedding: $x = sA(x)$

(B1) $a=b=c=0.1 \Longleftrightarrow x=(1,1,1)^T$

(B2) $a=b=c=0.5 \Longleftrightarrow x=(1,1,1)^T$

(B1)

Δs	0.25	0.30	0.45	0.50	0.55	0.60	0.75	1.00
Φ	38	36	33	26	25	27	32	33

(B2)

Δs	0.10	0.48	0.50	0.55	0.60	1.00
Φ	76	41	35	36	36	101

(Φ: number of iterations for all steps for $\varepsilon = 10^{-15}$)

All tests were carried out by Ms. M. Traunmüller on an IBM 360/44.

REFERENCES

[1] Avila, J.H., Continuation methods for nonlinear equations, Tech. Report TR 142, Univ. of Maryland, Computer Science Center (1971).

[2] Ben Israel, A., A Newton-Raphson method for the solution of systems of equations, J. Math. Anal. Appl. 15 (1966), 243-252.

[3] Deuflhard, P., A modified Newton method for the solution of ill-conditioned systems of nonlinear equations with applications to multiple shooting, Num. Math. 22 (1974), 289-315.

[4] Deuflhard, P., A stepsize control for continuation methods with special applications to multiple shooting techniques, Techn. Report TUM-Math. 7627, Univ. Munich, Inst. Math. (1976)

[5] Engl, H., Wacker, Hj., Zarzer, E., Stetigkeitsmethoden sowie Alternativen bei der Behandlung der singulären Situation, ISNM Vol. 38, Oberwolfacher Tagungsberichte (1977), 175-193.

[6] Drexler, F.J., Eine Methode zur Berechnung sämtlicher Lösungen von Polynomgleichungssystemen, Dissertation, Univ. Linz (1976), Num. Math. 29 (1977), 45-58.

[7] Ficken, F., The continuation method for functional equations, Comm. Pure Appl. Math. 4 (1951).

[8] Hackl, J., Wacker, Hj., Zarzer, E., Zulehner, W., Über die Aufspaltung eines singulären Operators beim Bifurkationsproblem, Univ. Linz, Inst. Math., Inst. Bericht 72 (1977).

[9] Keller, H.B., Numerical solution of bifurcation and nonlinear eigenvalueproblems, to appear in Appl. Bifurcation Theory, Academic Press, New York (1977).

[1o] Kantorovich, L.W., Akilow, G.P., Funktionalanalysis in normierten Räumen, Akademie Verlag Berlin (1964).

[11] Leder, D., Automatische Schrittweitensteuerung bei global konvergenten Einbettungsmethoden, ZAMM 54 (1974).

[12] Meyer, G., On solving nonlinear equations with a one-parameter imbedding, SIAM J. Num. Anal. 5 (1968).

[13] Ortega, J.M., Rheinboldt, W.C., Iterative solution of nonlinear equations in several variables. Academic Press, New York (197o).

[14] Rheinboldt, W.C., An adaptive continuation process for solving systems of nonlinear equations, Techn. Report TR 393, Univ. of Maryland (1975).

[15] Schwetlick, H., Ein neues Prinzip zur Konstruktion implementierbarer, global konvergenter Einbettungsalgorithmen, Beiträge zur Num. Math. 4 (1975), 215-228.

[16] Wacker, Hj., Eine Methode zur numerischen Lösung nichtlinearer Gleichungen Fredholmscher Art, ZAMM 49/11 (1969).

[17] Wacker, Hj., Ein Iterationsverfahren zur Lösung spezieller nichtlinearer Randwertprobleme, Computing 8 (1972), 275-291.

[18] Wacker, Hj., Feilmeier, M., Lory, P., Scheuring, H., DFG-Projekt Fe 125/1, Math. Inst., Univ. Linz, Institutsberichte 7,26 (1973-1975).

[19] Wacker, Hj., Minimierung des Rechenaufwandes für spezielle Iterationsverfahren vom Typ Minimales Residuum, Computing 18 (1977), 2o9-224.

[2o] Hackl, H., Wacker, Hj., Zulehner, W., Implementierbare Schrittweitensteuerungen bei globalisierten Iterationsverfahren, Univ. Linz, Math. Inst., Institutsbericht 1o1 (1978).

ON THE IMBEDDING METHOD FOR SEMILINEAR FIRST ORDER ELLIPTIC SYSTEMS AND RELATED FINITE ELEMENT METHODS

W. L. Wendland

Fachbereich Mathematik
Technische Hochschule Darmstadt
Darmstadt, Federal Republic of Germany

ABSTRACT

Based on a-priori estimates for the solutions of semi-linear first order elliptic boundary value problems in the plane, the Newton method with imbedding is investigated.

[1] This research was supported in part by the Fulbright Foundation through the Exchange-Visitor-Program G-I-5 and the Institute for Mathematical Sciences at the University of Delaware and the Oregon State University in 1977.

ISBN 0-12-729250-0

Then a finite element approximation in connection with the Newton imbedding is presented. It can be shown that the corresponding Newton imbedding for the approximates converges independently of the meshwidth h. Asymptotic error estimates of optimal order are obtained which generalize known results even for the linear problems to those including point conditions. The method combines the advantages which are known from Galerkin methods for second order elliptic problems as sparse discrete matrices with the treatment of a very big and general class of boundary conditions.

INTRODUCTION

Many problems in classical mathematical physics can be modeled by elliptic boundary value problems in the plane. For certain special cases, e. g. for the equations of elastic plates and shells, the characteristic roots of the principal part coinside and the equations can be transformed into a first order Douglis normal form [27, § 3.1, § 5.3] which have been studied first by Douglis in [10]. Here we consider Riemann Hilbert boundary value problems with semilinear equations in the Douglis normal form and linear boundary and side conditions to which we apply a finite element approximation. Both, the semilinear boundary value problem and its approximation are solved by the use of an imbedding method combined with the Newton procedure due to Wacker [24].

It is well known that such semilinear problems can be treated with the Leray-Schauder technique or Schauder's fixed point theorem [6, p. 282 ff.]. Besides compactness properties this appraoch demands a-priori bounds for the solutions and H. Schäfer [20] provided a very handy version of this kind of fixed point theorem.

For general nonlinear elliptic boundary value problems in connection with the Leray-Schauder theory there has been done already a vast amount of work (see e. g. L. Nirenberg [17]). But our problems have the beautiful advantage that the Bers-Haack-Vekua theory of generalized analytic functions can be applied. The problems considered here are slightly more general than those in [28]. Again, the approach is based on a-priori estimates for the solution. If they are available then by using suitable cut off functions the nonlinearity can be restricted to uniformly bounded terms. Hence, for the finite element approximation we always assume that the nonlinear function in the differential equations has compact support with respect to the unknowns. The finite element approximation is based on the least squares method for the corresponding *linear* problems, an extension of the method by Bramble and Schatz [9] to first order equations [27, chap. 8]. But there we did not handle the point conditions which appear here as side conditions. Hence, for linear problems, the results of [27, chap. 8] need first to be extended such that point condition are also admissible. The method for the nonlinear problems is some discrete version of the imbedding Newton method, an iterative procedure. For every iterate one has to solve a linear finite element problem related to the Ritz Galerkin method. It combines the advantages of Galerkin methods to second order problems with finite element functions, such as sparse matrices in the discrete equations, with the treatment of very general boundary and side conditions. It is shown that the discrete version of the imbedding Newton procedure converges uniformly for all $h \leq h_o$ where h denotes the meshwidth assigned to the finite element spaces in question. For the resulting finite element solution we find asymptotic error estimates of optimal order in the W_2^t Sobolev norms and also pointwise asymptotic error estimates, $t \geq 1$.

The problems in question and the corresponding imbedding Newton method are formulated in § 1 following [28]. In § 2 we present some sufficient conditions providing a-priori estimates for the solution. One of these conditions as well as the convergence results for the finite element approximations are based on a-priori estimates for linear problems with constants which are uniform for classes of linear equations having uniformly bounded coefficients of the terms without derivatives. Such estimates are needed in Hölder spaces as well as in spaces of generalized solutions in the W_p^1 Sobolev spaces. All these estimates are based on Vekua's theory of generalized analytic functions [23] and its generalization to hyperanalytic functions by Gilbert and Hile [12, 14] and corresponding boundary value problems [5].

In § 3 we investigate the least squares method going back to Bramble and Schatz [9] for linear first order problems with point conditions. The finite elements are restricted to regular elements providing best approximation and stability. The famous Courant elements belong to this class. Here we extend the results of [27, chap. 8] where point conditions are excluded to our case with point conditions.

In § 4 we formulate and investigate the imbedding Newton method for finite element approximations. Based on the estimates for linear problems we can prove the already mentioned convergence results for the imbedding Newton iteration and also asymptotic error estimates formulated in the main Theorems 4.3 and 4.4. The proofs are based on rather tedious uniform a-priori estimates to linear problems. Our convergence results contain as a special case Neumann problems with inhomogeneous boundary data and coincide for zero boundary values with those obtained by Höhn [15] and several other authors more recently. They are also related to the results for a rather wide class of nonlinear even order problems obtained by Nitsche [18]. A rather different approach is based on discrete Green and Neumann functions, [28, § 8.6] and corresponding error estimates which might imply even a more straightforward analysis than here (see also [3o]).

All the presented investigations are restricted to somewhat triangular systems of first order equations and boundary conditions which are solvable consecutively with generalized analytic functions. This provides uniqueness for every corresponding linearization and the nonlinear problems of the imbedded family. Hence, eigenvalues do never appear. This situation for our first order systems is similar to second order problems providing maximum principles or problems of positive type [7] [18]. Our boundary conditions are much more general than the usual Dirichlet or Neumann conditions.

For more general nonlinear first order systems the corresponding bifurcation problems are of great interest but this has yet to be done.

Numerical experiences with the proposed method are not yet available except those in [26] where only very simple problems were investigated with a method different from ours. But with the extremely good experiences with second order problems involving finite elements, the proposed imbedding Newton method should also lead to effective numerical schemes.

§ 1 BOUNDARY VALUE PROBLEMS FOR SOME SEMILINEAR ELLIPTIC SYSTEMS

Semilinear systems in the Douglis normal form are given by

$$(1.1)\quad Dw := w_{\bar z} + \begin{pmatrix} o & & & o \\ q_1 & \ddots & 0 & \\ \vdots & \ddots & \ddots & \\ q_{n-1} & \cdots & q_1 & o \end{pmatrix} w_z + Aw + B\bar w = g(z,w) + f(z)$$

where $w = \{w_o,\dots,w_{n-1}\}^T$ denotes the vector of n unknown complex valued functions. $q_1(z),\dots,q_{n-1}(z)$ are given coefficients as well as the matrices $A(z)$ and $B(z)$, $z = x + iy$. $g(z,w)$ is the given nonlinearity, a vector of n functions; $f(z) = \{f_o,\dots,f_{n-1}(z)\}^T$ denotes the given right hand side. Note that here all these functions are <u>not</u> supposed to be complex analytic, i. e. $g_{j\bar z} \neq o$, $A_{\bar z} \neq o$, $g_{\bar w} \neq o$ in general; i. e. we consider all functions as to depend on the <u>point</u> $z \in \mathbb{C}$, $w \in \mathbb{C}^n$ etc. rather than on the complex variables. The differentiations are defined as the usual Pompeju derivatives by

$$(1.2)\quad \partial/\partial\bar z = \tfrac{1}{2}(\partial/\partial x + i\partial/\partial y),\ \partial/\partial z = \tfrac{1}{2}(\partial/\partial x - i\partial/\partial y)$$

(see [23]). The form (1.1) can be represented more concisely by utilizing the algebra of hypercomplex numbers which can be identified with the commutative matrix algebra generated by the identity matrix

$$(1.3)\quad e^o := ((\delta_{jk})),\ j,k = o,\dots,n-1;\ \delta_{jk} \text{ the Kronecker symbol,}$$

the shifting matrix

$$(1.4)\quad e := ((\delta_{j,k+1})),\ j,k = o,n-1,$$

and the complex numbers. The elements of this algebra are called hypercomplex numbers. Following Douglis, the system (1.1) can be written as

$$(1.5) \quad Dw = w_{\bar{z}} + qw_z + \sum_{j=o}^{n-1} e^j \sum_{k=o}^{n-1} (A_{jk}\, w_k + B_{jk}\bar{w}_k) = g + f$$

with hypercomplex valued functions w,g,f,q etc. where

$$(1.6) \quad w = \sum_{j=o}^{n-1} e^j w_j, \; g = \sum_{j=o}^{n-1} e^j g_j, \; f = \sum_{j=o}^{n-1} e^j f_j \text{ and}$$

$$q = e \sum_{j=o}^{n-2} e^j q_j .$$

The theory of solutions to linear systems (1.5) with $g \equiv o$ is known as generalized hyperanalytic function theory and was far developed by Bojarski [8] and by Gilbert and Hile [12,14]. If, in addition, A and B and also the boundary conditions formed by lower triangular matrices then these linear problems are uniquely solvable regardless the size of coefficients or of the domain [5]. Thus we assume that (1.1) has the corresponding form and consider the following boundary value problems:

$$Dw = w_{\bar{z}} + qw_z + \sum_{j=o}^{n-1} e^j \sum_{k=o}^{j} (A_{jk}w_k + B_{jk}\bar{w}_k) = g(z,w) + f(z) \quad \text{in } G,$$

$$Rw := \sum_{j=o}^{n-1} e^j \operatorname{Re} \sum_{k=o}^{j} p_{jk}\, w_k = \psi := \sum_{j=o}^{n-1} e^j \psi_j \quad \text{on } \partial G$$

(1.7) $w_j(z_\ell) = \gamma_{j\ell},\ \ell = 1,\ldots,m_j,$ [1]

$$\oint_{\partial G} \sigma \ \mathrm{Im}\ \{w_j p_{jj}\}\ ds = \kappa_j,\ j = o,\ldots,n-1.$$

where the following assumptions will be made throughout this paper:

(A1) *G is a simple connected domain with smooth boundary curve* ∂G *in* $C^{1+\alpha}$, $\alpha > o$, i. e. *Hölder continuous tangent as function of the arclength*

$p_{jj} = \exp(i\tau_j)$ *on* ∂G *with winding numbers*

$$m_j = -\frac{1}{2\pi}\oint_{\partial G} d\tau_j \geq o.$$

The $z_\ell \in G$ *are fixed chosen points.* $\sigma > o$ *is a given smooth weight function.*

(A2) *All given functions are at least Hölder-continuous. Moreover, the nonlinear functions in* $g(z,w)$ *possess Gateaux derivatives* $(\frac{\partial}{\partial w_k} g)$ *and* $(\frac{\partial}{\partial \overline{w}_k} g)$ *in* $\overline{G} \times \mathbb{C}^{2n}$ *and*

$\|g\|_{C^\beta} \leq M'$ *for all* w *with* $\|w\|_{C^\beta} \leq M$
where M' *depends only on* M,

$\|g(z,w) - g(z,w')\|_{C^\beta},\ \|\partial_k g(z,w) - \partial_k g(z,w')\|_{C^\beta} \to o$

[1] *For* $m_j = o$ *it is understood here and in the following to skip the corresponding point condition.*

for $\|w-w'\|_{C^\beta} \to 0$ *where* ∂_k *denote the derivatives* $\partial/\partial w_k$ *and* $\partial/\partial \overline{w}_k$ *and for every* $0 \le \beta \le \alpha$.

The absolute values and norms are defined by

$$|\gamma|^2 = |\sum_{k=0}^{n-1} e^k \gamma_k|^2 := \sum_{k=0}^{n-1} |\gamma_k|^2 \text{ and}$$

$$\|w\| = \|\sum_{k=0}^{n-1} e^k w_k\| := \sum_{k=0}^{n-1} \|w_k\|$$

where $\|.\|_{C^\alpha}$ denote the usual Hölder norms or the supremum norm for $\alpha = 0$, $\|.\|_{L^p}$ denote L^p norms and $\|.\|_{W^s_p}$ the corresponding Sobolev space norms over $\overline{G}$. Sometimes we shall also use corresponding norms on ∂G which will be indicated by $\|.\|_{C^\alpha(\partial G)}$ etc.

For the nonlinear problem (1.7) we further demand :

(A3) *For any solution* w *of* (1.7) *there is an a-priori estimate available:*

(1.8) $\|w\|_{C^0} \le K < \infty$.

The latter assupmtion (1.8) allows us to replace the differential equations in (1.7) by

$$(1.9) \quad Dw = \sum_{k=0}^{n-1} e^k g_k(z,w)\, \chi_k(w) + f(z) = \tilde{g}(z,w) + f(z)$$

where $\chi_k(w)$ denote suitable C^∞ cut off functions with compact support and with $\chi_k \equiv 1$ for all w with $\|w\|_{C^0} \le K$.

Thus, from now on $\tilde{g}(z,w)$ will always be uniformly bounded for all $(z,w) \in \overline{G} \times \mathbb{C}^n$ having compact support with respect to w.

A simple conclusion is the existence of a solution to each $t \in [o,1]$ of the imbedded family of problems,

$$\begin{aligned} &Dw = t\tilde{g}(z,w) + f(z) \text{ in } G,\ t \in [o,1], \\ &\operatorname{Re} \sum_{k=o}^{j} p_{jk}(s) w_k = \psi_j(s) \text{ on } \partial G, \\ (1.10)\quad &w_j(z_\ell) = \gamma_\ell \quad \text{for } \ell = 1,\ldots,m_j, \\ &\oint_{\partial G} \sigma(s) \operatorname{Im} \{p_{jj} w_j\}\, ds = \kappa_j,\ j = o,\ldots,n-1 \end{aligned}$$

(see [20], [28]).

Thus, the a-priori estimate (1.8) is crucial for our imbedding as well as for later finite element approximations.

In § 2 we shall give some sufficient conditions on g in (1.7) in order to find a-priori bounds (1.8).

For the construction of the solution to (1.10) let us use a combination between the above imbedding and Newton's method which has been investigated by Wacker [24]. To this end we start with the solution w(o) of (1.10) subject to t=o, the linear boundary value problem. w(o) serves as the initial approximation for (1.10) with $t=t_1 > o$ for the Newton method. Since for Newton's method the iterate linear problems should be uniquely solvable, let us impose *an additional assumption on the nonlinear term* $\tilde{g}$ such that the corresponding linearized problems are always *of triangular shape*:

(A4) $$\frac{\partial \tilde{g}_\ell}{\partial w_k} = o \quad \text{and} \quad \frac{\partial \tilde{g}_\ell}{\partial \overline{w}_k} = o \quad \text{for } k > \ell.$$

Note, that these derivatives are uniformly bounded.

For a fixed value of $t = t_j$, the Newton method in connection with (1.10) reads as

$$Dw^{(m+1)} = t_j \sum_{\ell=o}^{n-1} e^\ell \Big[\sum_{k=o}^{\ell} \Big(\frac{\partial \tilde{g}_\ell}{\partial w_k}(z,w^{(m)}) \{w_k^{(m+1)} - w_k^{(m)}\}$$

$$+ \frac{\partial \tilde{g}_\ell}{\partial \overline{w}_k}(z,w^{(m)}) \{\overline{w}_k^{(m+1)} - \overline{w}_k^{(m)}\}\Big]$$

$$+ t_j \tilde{g}(z,w^{(m)}) + f,$$

(1.11)

$$\text{Re} \sum_{k=o}^{j} p_{jk}(s) w_k^{(m+1)} = \psi_j(s) \quad \text{on } \partial G,$$

$$w_j^{(m+1)}(z_\ell) = \gamma_{j\ell} \quad \text{for } \ell = 1,\ldots,m_j,$$

$$\oint_{\partial G} \sigma(s) \text{Im}\{p_{jj} w_j^{(m+1)}\} ds = \kappa_j, \quad j = o,\ldots,n-1;$$

$$w^{(o)} := w(z,t_{j-1}).$$

Clearly, this is an iteration where $w^{(m+1)}$ is a solution of a *linear* problem containing the already known $w^{(m)}$. Its limit defines $w(z,t_j)$ which serves as the initial value for the next iteration with $t = t_{j+1}$. Since it is just the initial step for the next iteration, in practice it has to be replaced by a suitable iterate $w^{(m)}(z,t_{j-1})$ for a finite m being large enough. The convergence of the method is assured by the following two theorems.

Theorem 1.1: Under the above assumptions the equations (1.11) *are uniquely solvable for every* $t_j \in [o,1]$ *and each* $m=1,2,\ldots$. *The Theorem remains valid for weak solutions*

$$w^{(m+1)} \in W_p^1,\ p > 2 \quad \text{if}\ \frac{\partial \tilde{g}_\ell}{\partial w_k},\ \frac{\partial \tilde{g}_\ell}{\partial \overline{w}_k},\ \tilde{g},\ f \in L^p \quad \textit{and}$$

$$\psi_j \in C^{\frac{2-p}{p}}(\partial G) \cap W_p^{1-\frac{1}{p}}(\partial G).$$

Proof: Let us consider the case of $C^{1+\alpha}$ solutions first. We observe that for every t_j and m, the linear equations in (1.11) are of the form (1.5) with $\tilde{g} \equiv o$ but coefficients A_{jk} and B_{jk} replaced by $A_{jk} - \frac{\partial \tilde{g}_j}{\partial w_k}(z,w^{(m)})$ and $B_{jk} - \frac{\partial \tilde{g}_j}{\partial \overline{w}_k}(z,w^{(m)})$ respectively and corresponding $f^{(m)}$.

Such triangular system together with the triangular boundary conditions can be solved recursively beginning with the first equation for $w_j^{(m+1)}$ with j=o and the corresponding boundary and side conditions. This is just a standard problem for $w_o^{(m+1)}$ possessing a unique solution in $C^{1+\alpha}(\overline{G})$ (see [27, Theorem 1.2.4]). Inserting $w_o^{(m+1)}$ into the next equation for $w_1^{(m+1)}$ and the boundary and side conditions in (1.11) with j=1 we obtain again a boundary value problem of the standard type for $w_1^{(m+1)}$ yielding a unique solution. Repeating this procedure, we finally find $w_{n-1}^{(m+1)}$ and, thus, $w^{(m+1)}$.

For the construction of weak solutions we proceed in the same way using Vekua's results [23] and [8], [14]. Note that in this case $w^{(m+1)} \in C^{\frac{p-2}{p}}(\overline{G})$ and that therefore the point conditions define continuous linear functionals on the corresponding space of solutions.

For the convergence proof we require stronger assumptions for the nonlinear terms.

(A5) *The second Gateaux derivatives of* $\tilde{g}(z,w)$ *with respect to* w *exist in* $\overline{G} \times \mathbb{C}^{2n}$, i. e.

$$\frac{\partial^2 \tilde{g}}{\partial w_k \partial w_j}, \quad \frac{\partial^2 \tilde{g}}{\partial \overline{w}_k \partial w_j}, \quad \frac{\partial^2 \tilde{g}}{\partial \overline{w}_k \partial \overline{w}_j}$$

defining bounded continuous mappings in $C^\beta(\overline{G})$ *for every* β *with* $o \le \beta \le \alpha$.

Assuming (A5) we have

Theorem 1.2: *There exists a finite partition* $t_o = o < t_1 < t_2 < \ldots < t_N = 1$ *such that the imbedding Newton method converges.*

Proof: For a fixed $t=t_j$, the difference of two iterates of (1.11) satisfies the system

$$D(w^{(m+1)}-w^{(m)}) = t_j \sum_{\ell=o}^{n-1} e^{\ell} \sum_{k=o}^{\ell} [\frac{\partial\tilde{g}_\ell}{\partial w_k}(z,w^{(m)})\{w_k^{(m+1)}-w_k^{(m)}\}$$

$$+ \frac{\partial\tilde{g}_\ell}{\partial\overline{w}_k}(z,w^{(m)})\{\overline{w}_k^{(m+1)} - \overline{w}_k^{(m)}\}$$

$$(1.12) \qquad + \frac{1}{2}\sum_{p=o}^{\ell} (\tilde{g}_{\ell kp11}\{w_k^{(m)} - w_k^{(m-1)}\}\{w_p^{(m)} - w_p^{(m-1)}\}$$

$$+ 2\tilde{g}_{\ell kp12}\{w_k^{(m)} - w_k^{(m-1)}\}\{\overline{w}_p^{(m)} - \overline{w}_p^{(m-1)}\}$$

$$+ \tilde{g}_{\ell kp22}\{\overline{w}_k^{(m)} - \overline{w}_k^{(m-1)}\}\{\overline{w}_p^{(m)} - \overline{w}_p^{(m-1)}\})],$$

$$\mathrm{Re}\sum_{k=o}^{j} p_{jk}(s)(w_k^{(m+1)} - w_k^{(m)}) = o \quad \text{on } \partial G,$$

$$(1.13) \qquad (w_j^{(m+1)} - w_j^{(m)})(z_\ell) = o \quad \text{for } \ell = 1,\ldots,m_j,$$

$$\oint_{\partial G} \sigma(s)\ \mathrm{Im}\ \{p_{jj}(w_j^{(m+1)} - w_j^{(m)})\}\ ds = o \qquad j=o,\ldots,n-1,$$

where the $\tilde{g}_{\ell kp\alpha\beta}$ are defined recursively from (1.11) by using Taylor's formula; i. e.

$$(1.14) \quad \tilde{g}_{\ell kp11} \equiv \int_o^1 \frac{\partial^2\tilde{g}_\ell}{\partial w_k \partial w_p}(z,\ \tau w^{(m)} + (1-\tau)w^{(m-1)})d\tau$$

and similar formulas for the remaining terms. Now (1.12) - (1.14) and our assumptions (A1-A5) yield an a-priori estimate which will be proved in § 2, (2.9),

$$(1.15) \quad \| w^{(m+1)} - w^{(m)} \|_\alpha \le Kt_j \| w^{(m)} - w^{(m-1)} \|_\alpha^2$$

where K is *independent* of $t_j \varepsilon [o,1]$ and $w^{(m)}$, $w^{(m-1)}$.

Thus for every t_j, the Newton iteration converges if $w^{(o)}(t_j) = w(t_{j-1})$ is chosen such that

(1.16) $Kt_j \; \| w^{(1)} - w^{(o)} \|_\alpha < 1.$

Hence, we consider

$$D(w^{(1)} - w^{(o)}) = t_j \sum_{\ell=o}^{n-1} e^\ell \sum_{k=o}^{\ell} \frac{\partial g_\ell}{\partial w_k}(z,w^{(o)})\{w_k^{(1)} - w_k^{(o)}\}$$

$$(1.17) \qquad \frac{\partial \tilde{g}_\ell}{\partial \overline{w}_k}(z,w^{(o)}) \; \{\overline{w}_k^{(1)} - \overline{w}_k^{(o)}\}$$

$$+ (t_j - t_{j-1})\tilde{g}(z,w^{(o)})$$

together with the corresponding homogeneous boundary and point conditions (1.13).

Using the a-priori estimate (2.9) for $w^{(1)} - w^{(o)}$ we find

$$(1.18) \quad \| w^{(1)} - w^{(o)} \|_\alpha \le \gamma(t_j - t_{j-1}) \| \tilde{g}(.,w^{(o)}) \|_\alpha$$

$$\le \tilde{K}(t_j - t_{j-1})$$

where $\tilde{K}$ is independent of t_j and $w^{(o)}$ according to (A2), (A3), the compact support of $\tilde{g}$ with respect to w.

Hence, (1.16) is satisfied if we choose t_j such that

(1.19) $t_j(t_j - t_{j-1}) < (K\tilde{K})^{-1}.$

Since the right hand side is independent of t_j, the desired choice of finitely many $t_o = o < t_1 < \dots < t_k = 1$ obeying (1.19) is always possible.

§ 2 A-PRIORI ESTIMATES

The basis of Schäfer's version of the Schauder-Leray technique as well as of the imbedding method is an a-priori estimate for all solutions of (1.10) and the corresponding linear problems where $\tilde{g} \equiv o$. For the latter everybody knows the Agmon-Douglis-Nirenberg estimates up to the boundary. Here we shall use in addition slightly modified versions involving bounds of the coefficients in (1.7) more explicitely. Based on these estimates we repeat the ideas of [28] and [4], [13] obtaining sufficient conditions for a-priori estimates (1.8) of the solutions to the nonlinear problem (1.7). If g satisfies Lipschitz conditions with small enough Lipschitz constants then by using the contraction principle one can also show a-priori bounds for systems which have been investigated by Tutschke [21], [22].

For the desired a-priori estimates let us begin with refering the well known Agmon-Douglis-Nirenberg estimates for linear problems corresponding to Theorem 1.1.

Theorem 2.1: *Let* $g \equiv o$ *in* (1.7). *Then for every solution of* (1.7) *there hold the following a-priori estimates:*

$$(2.1)\quad \|w\|_{C^{1+\alpha}} \le k_1\{\|f\|_{C^\alpha} + \|\psi\|_{C^{1+\alpha}(\partial G)} + \sum_{j=o}^{n-1}(|\kappa_j| + \sum_{\ell=1}^{m_j}|\gamma_{j\ell}|)\},$$

$$(2.2)\quad \|w\|_{C^\gamma} \le \gamma_1\|w\|_{W_p^1} \le k_2\{\|f\|_{L^p} + \|\psi\|_{C^\gamma(\partial G)} + \|\psi\|_{W_p^{1-\frac{1}{p}}(\partial G)} + \sum_{j=o}^{n-1}(|\kappa_j| + \sum_{\ell=1}^{m_j}|\gamma_{j\ell}|),$$

$$(2.3)\quad \|w\|_{W_2^s} \le k_3\{\|f\|_{W_2^{s-1}} + \|\psi\|_{W_2^{s-\frac{1}{2}}(\partial G)} + \sum_{j=o}^{n-1}(|\kappa_j| + \sum_{\ell=1}^{m_j}|\gamma_{j\ell}|)\}$$

where $p > 2, \gamma = \frac{p-2}{p}$, $s > 1$ *and the constants* γ_1, k_1, k_2, k_3 *depend on* $\alpha, \gamma, s, G, q, A_{jk}, B_{jk}, p_{jk}, \sigma$ *and* $z_\ell \in G, o < \alpha < 1$.

The estimates (2.1), (2.3) are well known, see e. g. [27, chap. 4], (2.2) follows from Vekua's results in [23], S. 182, Satz 4.1 and S. 233, Satz 4.13 and a slight modification of the proofs by using Satz 1.37, S. 62 and the Agmon-Douglis-Nirenberg estimates; see also [14] .

Since the a-priori estimates (2.3) for $s \le 1$ are not available for problems with point conditions we have to consider auxiliary linear problems of the form

(2.4) $$Dw = f$$

$$Rw = \sum_{j=o}^{n-1} e^j \operatorname{Re} \sum_{k=o}^{j} p_{jk} w_k = \sum_{j=o}^{n-1} e^j \psi_j = \psi \quad \text{on } \partial G$$

(2.5)

$$\Lambda_\ell \; w = \gamma_\ell \in \mathbb{R}, \; \ell = 1,\ldots,N = n + \sum_{j=o}^{n-1} m_j$$

where the Λ_ℓ are real linear functionals. Recall that all functions are considered in spaces over the reals. We assume:

(A6) *The* Λ_ℓ *are bounded on* $W_2^o = L^2$ *and the problem* (2.4) *is uniquely solvable.*

Then we have the following sharper Agmon-Douglis-Nirenberg estimate proved by Dikanski:

Theorem 2.2 [27, *Lemma 8.2.1*]: *Let* $w \in W_2^t \cap W_2^1$, $t \geq o$, *be a generalized solution of* (2.4). *Then we have*

(2.6) $$\|w\|_{W_2^t} \leq k_4\{\|f\|_{W_2^{t-1}} + \|\psi\|_{W_2^{t-\frac{1}{2}}(\partial G)} + \sum_{\ell=1}^{N} |\Lambda_\ell w|\}$$

As we have already indicated, the convergence results for the imbedding method hinge on a-priori estimates which are *uniform* for classes of equations.

Theorem 2.3: For all $h = \sum_{k=o}^{n-1} e^k h_k$ *satisfying the homogeneous conditions*

$$(2.7) \quad \mathrm{Re} \sum_{k=o}^{j} p_{jk}(s) h_k = o \quad \text{on} \quad \partial G,$$

$$h_j(z_\ell) = o \quad \text{for} \quad \ell = 1, \ldots, m_j,$$

$$\oint_{\partial G} \sigma(s) \ \mathrm{Im} \{p_{jj}(s) h_j\} \, ds = o, \ j=o, \ldots, n-1$$

there hold the a-priori estimates

$$(2.8) \quad \|h\|_{C^{1+\alpha}} \le k' \|Dh\|_{C^\alpha} \cdot \exp(k \cdot \max\{\|A_{jk}\|_{C^\alpha} + \|B_{jk}\|_{C^\alpha}\})$$

for $\alpha > o$ *and*

$$(2.9) \quad \|h\|_{C^\gamma} \le \gamma_1 \|h\|_{W^1_p} \le k' \|Dh\|_{L^p} \exp(k_1 \max\{\|A_{jk}\|_{L^p} + \|B_{jk}\|_{L^p}\})$$

for $p > 2$, $\gamma = \frac{p-2}{p}$, *where the constants* γ_1, k, k' *are independent of* h, A_{jk}, B_{jk}, *and where* $h \in C^{1+\alpha}$ *and* A_{jk}, $B_{jk} \in C^\alpha$ *in (2.8) and* $h \in W^1_p$, A_{jk}, $B_{jk} \in L^p$ *in (2.9).*

Proof: We proceed similarly to [28]. We shall use the same letter C for different constants in estimates if the distinction is not necessary.

In the special case $A_{jk} = B_{jk} \equiv o$ the estimates (2.8), (2.9) follow directly from (2.1), (2.2) with (2.7) and $Dh = f$. For the more general case we use the triangular shape of D and $((p_{jk}))$ in (1.10). Let us begin with the estimates for h_o. To this end let v be the solution of

$$\text{(2.10)} \quad v_{\bar z} = \begin{cases} -A_{oo} - B_{oo}\ \bar h_o \big/ h_o & \text{where } h_o \neq o, \\ -A_{oo} & \text{where } h_o = o \text{ and} \end{cases}$$

$$\operatorname{Im} v = o \text{ on } \partial G, \quad \oint_{\partial G} \operatorname{Re} v\ \sigma\, ds = o.$$

Then (2.9) implies with $m_o = o$ and $p_{oo} \equiv i$ the estimate

$$\text{(2.11)} \quad \| v \|_{C^\gamma} \le \gamma_1 \| v \|_{W^1_p} \le k(\| A_{oo} \|_{L^p} + \| B_{oo} \|_{L^p}).$$

Setting

$$\text{(2.12)} \qquad f \equiv h_o \exp(-v) - \beta\phi_o$$

where ϕ_o is holomorphic, defined by

$$\phi_{o\bar z} = o, \ \operatorname{Re}\phi_o e^{i\tau_o} = o \text{ on } \partial G, \quad \oint_{\partial G} \operatorname{Im}\phi_o\ e^{i\tau_o}\, ds = 1,$$

$$\phi_o(z_\ell) = o, \ \ell = 1,\dots,m_o,$$

where $p_{oo} = e^{i\tau_o}$ and β is choosen as

$$\text{(2.13)} \qquad \beta \equiv \oint_{\partial G} \sigma \operatorname{Im} h_o e^{i\tau_o - v}\, ds,$$

one obtains with

$$(2.14)\quad f_{\bar z} = e^{-V}\,(h_{o\bar z} + A_{oo}h + B_{oo}\bar h_o)$$

and $\quad \mathrm{Re}\, f\, e^{i\tau_o} = o$ on ∂G, $\oint_{\partial G} \mathrm{Im}\, f\, e^{i\tau_o}\, ds = o$, $f(z_\ell) = o$

from (2.2) the estimate

$$(2.15)\quad \|f\|_{C^\gamma} \le \gamma_1 \|f\|_{W_p^1} \le k'\,\|h_{o\bar z} + A_{oo}h_o + B_{oo}\bar h_o\|_{L^p}\|e^{-V}\|_{C^o}$$

$$\le k' \exp(k(\|A_{oo}\|_{L^p} + \|B_{oo}\|_{L^p}))\cdot$$

$$\cdot\,\|h_{o\bar z} + A_{oo}h_o + B_{oo}\bar h_o\|_{L^p}.$$

Since $\|f\cdot g\|_{C^\gamma} \le \|f\|_{C^\gamma}\|g\|_{C^\gamma}$ and

$$\|f\cdot g\|_{W_p^1} \le c\,\|f\|_{W_p^1}\cdot\|g\|_{W_p^1}$$

for $p > 2$ we find from (2.12) with (2.11) and (2.15) the estimate

$$\|h_o\|_{C^\gamma} \le \gamma_1\|h_o\|_{W_p^1} \le c\,\|e^V\|_{W_p^1}\{\|f\|_{W_p^1} + |\beta|\;\|\phi_o\|_{W_p^1}\}$$

$$(2.16)\quad \le \exp(k(\|A_{oo}\|_{L^p} + \|B_{oo}\|_{L^p}))\cdot$$

$$\cdot\{k'\exp(k(\|A_{oo}\|_{L^p}+\|B_{oo}\|_{L^p}))c\{\|h_{o\bar z}+A_{oo}h_o+B_{oo}\bar h_o\|_{L^p}+k'|\beta|\}.$$

Now it only remains to estimate $|\beta|$. Here we use (2.12) and (2.7) obtaining

$$(2.17)\quad \beta = - \oint_{\partial G} \sigma \operatorname{Im} f\, e^{i\tau_o + v} ds \cdot \{ \oint_{\partial G} \operatorname{Im} \phi_o\, e^{i\tau_o + v} ds\}^{-1}$$

where the nominator never vanishes according to [15, § 11.1]. In (2.17) we use again (2.15) and (2.16) and find

$$(2.18)\ |\beta| \le k'' \exp(3k(\|A_{oo}\|_{L^p} + \|B_{oo}\|_{L^p}) \cdot \|h_{o\bar{z}} + A_{oo}h_o + B_{oo}\bar{h}_{oo}\|_{L^p}.$$

Inserting (2.18) into (2.16) we find the first component in (2.9).

If $h \in C^{1+\alpha}$ and the coefficients are Hölder continuous then we begin with choosing p big enough in (2.11) such that $\gamma \ge \alpha$. Then for f we use the a-priori estimate (2.1) replacing (2.15) by the corresponding estimate of $\|f\|_{C^{1+\alpha}}$ and, similarly for $\|h\|_{C^{1+\alpha}}$ in (2.16). With the new estimates in (2.17) we obtain Hölder norms in (2.18) and, finally, the desired estimate (2.8) for the first component.

The next component h_1 of h satisfies the conditions

$$(2.19)\quad \begin{aligned} &\operatorname{Re} e^{i\tau_1} h_1 = \operatorname{Re} h_1 p_{11} = -\operatorname{Re} h_o\, p_{1o} \quad \text{on } \partial G, \\ &h_1(z_\ell) = o \quad \text{for } \ell = 1,\dots,m_1 \quad \text{and} \\ &\oint_{\partial G} \sigma(s) \operatorname{Im} p_{11}(s) h_1(s) ds = o. \end{aligned}$$

Since the first condition is inhomogeneous for h_1, we decompose h_1 into

$$(2.20) \quad h_1 = h_1' + \tilde{h}_1$$

where h_1' is a *holomorphic function* satisfying (2.19). Thus we have from (2.2) an estimate

$$(2.21) \quad \begin{aligned} \| h_1' \|_{C^\gamma} \le \gamma_1 \| h_1' \|_{W_p^1} &\le k_2 \{ \| \operatorname{Re} h_o p_{1o} \|_{C^\gamma(\partial G)} + \\ &+ \| \operatorname{Re} h_o p_{1o} \|_{W_p^{1-\frac{1}{p}}(\partial G)} \} \le k_2' \| h_o \|_{W_p^1} \end{aligned}$$

with a suitable constant k_2' and the corresponding estimate

$$(2.22) \quad \| h_1' \|_{C^{1+\alpha}} \le k_1' \| h_o \|_{C^{1+\alpha}}$$

for Hölder continuous coefficients. For the remaining function $\tilde{h}_1$ we find from

$$\begin{aligned} \operatorname{Re} p_{11} \tilde{h}_1 &= o \qquad \text{on } \partial G, \\ \tilde{h}_1(z_\ell) &= o \qquad \text{for } \ell = 1,\dots,m_1 \text{ and} \\ \oint_{\partial G} \operatorname{Im} p_{11} \tilde{h}_1 \, ds &= o \end{aligned}$$

in the same way as above for h_o, an estimate for h_1:

$$(2.23) \quad \begin{aligned} \| \tilde{h}_1 \|_{C^\gamma} \le \gamma_1 \| \tilde{h}_1 \|_{W_p^1} &\le \tilde{\gamma}_1 \| \tilde{h}_{1\bar{z}} + A_{11} \tilde{h}_1 + B_{11} \bar{\tilde{h}}_1 \|_{L^p} \cdot \\ &\cdot \exp\{ \tilde{\gamma}_1' (\| A_{11} \|_{L^p} + \| B_{11} \|_{L^p}) \} \end{aligned}$$

and the corresponding estimate for $\| \tilde{h}_1 \|_{C^{1+\alpha}}$ if $A_{11}, B_{11} \in C^\alpha$ and $h \in C^{1+\alpha}$.

Using triangle inequality and (2.9), (2.21), (2.23) we finally find the desired estimate for h_1:

$$\|h_1\|_{C^\gamma} \leq \gamma_1 \|h_1\|_{W_p^1} \leq k_1 \|h_{1\bar{z}} + q_1 h_{oz} + A_{1o} h_o + A_{11} h_1$$

$$(2.24) \qquad + B_{1o}\bar{h}_o + B_{11}\bar{h}_1\|_{L^p} \cdot$$

$$\cdot \exp\{k_1' \max_{o \leq j,k \leq 1} \{\|A_{jk}\|_{L^p} + \|B_{jk}\|_{L^p}\}\}.$$

Correspondingly, one gets the estimate for $\|h_1\|_{C^{1+\alpha}}$.

Proceeding in the same way for $h_2, \ldots, h_{n-1}$, one finally finds (2.8) and (2.9).

Now let us formulate some sufficient conditions yielding an a-priori estimate (A.3). The first of these estimates is based on a growth condition for $g(z,w)$ in connection with Gronwall's inequality.

Theorem 2.4: Let us assume that for g *in* (1.7) *besides* (A.2), (A.4) *there exist two continuous monotonic increasing functions* $\omega(\rho)$, $\omega'(\rho)$ *for* $\rho \geq o$ *with* $\omega(o) = \omega'(o) = o$ *such that*

$$(2.25) \quad \|g\|_{C^\alpha} \leq \omega(\|w\|_{C^\alpha}) \quad \text{and} \quad \|\frac{\partial g}{\partial w_k}\|_{C^\alpha} + \|\frac{\partial g}{\partial \bar{w}_k}\|_{C^\alpha} \leq \omega'(\|w\|_{C^\alpha})$$

hold for all $w \in C^\alpha(\bar{G})$. *Let* $w(o)$ *denote the solution of* (1.10) *with* $t = o$ (which exists due to Theorem 1.1). *Let as further denote by* $\rho(t)$ *the solution of the Volterra equation*

$$\rho(t) = k'\omega \; (\|w(o)\|_{C^\alpha} + \int_p^t \rho(\tau)d\tau) \;\cdot$$

$$(2.26) \qquad \cdot \; \exp(k\{\max\{\|A_{jk}\|_{C^\alpha}+\|B_{jk}\|_{C^\alpha}\}+t\omega'(\|w(o)\|_{C^\alpha}+\int_o^t \rho(\tau)d\tau)\})$$

in $o \leq t < T$.

Then there exists exactly one solution of the boundary value problem

$$Dw = tg(z,w) + f \qquad \text{in } G$$

$$(2.27) \quad Rw = \sum_{j=o}^{n-1} e^j \,\mathrm{Re} \sum_{k=o}^{j} p_{jk}(s)w_k = \psi \quad \text{on } \partial G$$

$$w_j(z_\ell) = \gamma_{j\ell} \quad \text{for } \ell = 1,\ldots,m_j,$$

$$\oint_{\partial G} \sigma \mathrm{Im}(p_{jj}w_j)ds = \kappa_j \quad \text{for } j = o,1,\ldots,n-1$$

to every t *with* $o \leq t < T$ *satisfying the a-priori estimate*

$$(2.28) \qquad \|w\|_{C^\alpha} \leq \|w(o)\|_{C^\alpha} + \int_o^t \rho(\tau)d\tau.$$

Proof: We essentially repeat the arguments of [28] and consider the family of problems (2.27) with the solution $w(t)$ depending continuously differentiable on t. We can write

$$(2.29) \qquad w(t) = w(o) + \int_o^t \dot{w}(\tau)d\tau$$

where w(t) satisfies the Davidenko differential equation to (2.27) i. e. the boundary value problem

$$D\dot{w} = g(z,w(t)) + t \sum_{\ell=o}^{n-1} e \sum_{k=o}^{\ell} \{\frac{\partial g_\ell}{\partial w_k}(z,w(t)) \cdot \dot{w}_k + \frac{\partial g_\ell}{\partial \overline{w}_k}(z,w(t)) \cdot \overline{\dot{w}}_k\}$$

$$\text{Re} \sum_{k=o}^{j} p_{jk} \dot{w}_k = o \qquad \text{on } \partial G,$$

(2.30) $$\dot{w}_j(z_\ell) = o \qquad \text{for } \ell = 1,\ldots,m_j,$$

$$\oint_{\partial G} \sigma \text{Im}(p_{jj}\dot{w}_j)ds = o \text{ for } j = o,\ldots,n-1.$$

Hence, the a-priori estimate (2.8) can be applied for $\dot{w}$, i. e.

(2.31) $$||\dot{w}||_{C^\alpha} \le ||\dot{w}||_{C^{1+\alpha}} \le k'||g(\cdot,w(t))||_{C^\alpha} \cdot \exp\{k\{\max_{j,k}\{||A_{jk}||_{C^\alpha} + ||B_{jk}||_{C^\alpha} + t\,||\frac{\partial g_j}{\partial w_k}(\cdot,w(t))||_{C^\alpha} + t\,||\frac{\partial g_j}{\partial \overline{w}_k}(\cdot,w(t))||_{C^\alpha}\}\}.$$

Inserting

$$||w(t)||_{C^\alpha} \le ||w(o)||_{C^\alpha} + \int_o^t ||\dot{w}(\tau)||_{C^\alpha}\, d\tau$$

from (2.29) together with (2.25) into (2.31) we find

$$(2.32)\quad \begin{aligned} ||\dot{w}||_{C^\alpha} \le\ & k'\omega(||w(o)||_{C^\alpha} + \int_o^t ||\dot{w}||_{C^\alpha}\, d\tau)\cdot \\ & \cdot\exp(k\{\max\{||A_{jk}||_{C^\alpha}+||B_{jk}||_{C^\alpha}\}+t\omega'(||w(o)||_{C^\alpha}+\int_o^t ||\dot{w}||_{C^\alpha} d\tau)\}) \end{aligned}$$

Using the Gronwall inequality [25, p. 16] we find the desired a-priori estimate (2.28) for any solution of (2.27). Now Schäfer's theorem [20] yields existence and the triangular form of (2.30) yields uniqueness due to Theorem 1.1.

Another a-priori estimate involving more explicit bounds can be found in [13], [4]. There the more special problem

$$(2.33)\quad \begin{aligned} w_{\bar{z}} + qw_z &= g(z,w) && \text{in } G, \\ \text{Re } w &= o && \text{on } \partial G, \\ \oint_{\partial G} \text{Im } w|\phi'|ds &= o \end{aligned}$$

has been investigated where $\varphi(z)$ denotes the conformal mapping of G onto the unit disk. There it is assumed that:

(A7) $g(z,w)$ *is defined in* $\overline{G} \times \mathbb{C}^n$,

$g(z,w(z)) \in L^p$ *for every* $w \in C^o(\overline{G})$ *with some* $p > 2$,

$||g(z,w)-g(z,v)||_{C^o} \le k\omega(z,|w-v|)$

where $\omega(z,\rho)$ *with* $\omega(z,o) = o$ *is monotonic increasing with* ρ *and* $\omega(\cdot,v(.)) \in L^p$ *for every* $o \le v \in C^o$. *There exists a constant* K *such that*

$$\int_G\int \omega(z,K)\,\frac{dxdy}{|\zeta-z|} \le K.$$

Let π *denote the Green operator* [13, (2.9)], *a composition of conformal mappings and the two dimensional Hilbert transformation mapping* C^γ *into itself continuously,* $\gamma = \frac{p-2}{p}$, *let*

$$M := (n-1) \cdot \|q\|_{C^\gamma} \|\pi\|_{C^\gamma, C^\gamma}$$

and
$$c := \max_{\overline{G}} \left| \frac{\phi'(z)(\zeta - z)}{\phi(\zeta) - \phi(z)} \right| .$$

Then let the constant K *satisfy the inequality*

$$\left(\frac{2\pi(p-1)}{p-2}\right)^{1/d} d^\gamma \{8c \frac{M^n - 1}{M-1} \|g(z,o)\|_{L^p} + \|\omega(z,2K)\|_{L^p}\} \le K,$$

d = diam (G).

Theorem 2.5 [*13*]*: If the assumptions* (A7) *are fulfilled then for every solution of* (2.33) *there holds the a-priori estimate*

$$\|w\|_{C^o} \le K.$$

According to Gilbert [13], the assumptions (A7) can always be satisfied if d is small enough, i. e. if the domain is small enough for a given differential equation. This corresponds to a small Lipschitz constant of g with respect to w and to Tutschke's assumptions in [21], [22].

§ 3 THE LEAST SQUARES FINITE ELEMENT APPROXIMATION FOR LINEAR PROBLEMS

In the following let $\tilde{H}(h)$ be a sequence of regular real finite element spaces $\tilde{H} \cap W_2^m(G)$, $m \geq 1$, m an integer, defined over the region $\overline{G}$. For simplicity, let us use triangular elements following Fix and Strang [11]. Let $\overline{G}$ be covered by a regular family of triangular nets, i. e. if $h > o$ denotes the maximal side length of the triangles of one of these nets then we require that every triangle of the particular net contains a disk of radius $c \cdot h$ where c is a constant independent of h, $o < c < \frac{1}{2}$. The family corresponds to a sequence in h with $h \to o$. We further assume that each triangle T containing points of ∂G allows the placement of a disk with radius $c \cdot h$ into the intersection $\overline{T} \cap \overline{G}$. The choice of the integer m depends on the smoothness and accuracy desired. To each of the triangulations we define $\tilde{H}(h)$ to be the space of piecewise real polynomials in x and y of fixed degree $\tilde{m}$. Moreover, the additional assumption $\tilde{H}(h) \subset W_2^m \cap C^{m-1}$ is required restricting the possible $\tilde{m}$. E. g. for $m = 1$, the piecewise linear functions form the famous Courant elements. We further assume that $\tilde{H}(h)$ possesses a basis of trial functions $\chi_1, \ldots, \chi_{M(h)}$ satisfying a uniformity condition of order m implying with the uniform triangulations the following properties of convergence (C) and stability (S):

(C) *For* $k \le r$ *with* $-m-1 \le k \le m$, $-m \le r \le m+1$ *and for any* $v \in W_2^r$ *there exists an* $V \in \tilde{H}$ *such that*

$$(3.1)\qquad \|V - v\|_{W_2^k} \le c_{rk}\, h^{r-k}\, \|v\|_{W_2^r}$$

holds where V *is independent of* k *and* c_{rk} *is independent of* V, v *and* h.

(S) *For* $k \le r$ *with* $|r|, |k| \le m$ *there exists* $M_{rk} > 0$ *independent of* h *and* $\chi \in \tilde{H}$ *such that*

$$(3.2)\qquad \|\chi\|_{W_2^r} \le M_{rk}\, h^{k-r}\, \|\chi\|_{W_2^k} \quad \textit{for all } \chi \in \tilde{H}.$$

For such finite element spaces see e. g. [2], [3], [9], [11], [18], [19]. Since we shall also need pointwise estimates, (C) and (S) imply the following further stability and point estimates:

Lemma 3.1 [*28, Lemma 8.3.3*], [*18*]: *For* $m \ge 1$ *there exists a constant* M *independent of* h *and* $\chi \in \tilde{H}$ *such that*

$$(3.3)\qquad \|\chi\|_{C^0} \le M|\log h|^{\frac{1}{2}}\, \|\chi\|_{W_2^1} \quad \textit{for all } \chi \in \tilde{H}.$$

Lemma 3.2: *For* $v \in W_2^s \cap C^0$, $1 \le s \le m+1$, *the approximation* V *of* (C) *also satisfies the pointwise estimate*

$$(3.4)\quad |V(z_o)-v(z_o)| \leq c|\log h|^{\frac{1}{2}}\, h^{s-1}\, \|v\|_{W_2^s}\ ,\ z_o \in \overline{G},$$

where c *is independent of* h, v *and* z_o.

Proof: Let Λ_h denote the linear functional

$$(3.5)\quad \Lambda_h\, u := \frac{1}{4h^2} \iint\limits_{\substack{|x-x_o|\leq h\\ |y-y_o|\leq h}} u[dx,dx].$$ [2]

Then the Taylor formula and the Schwarz inequality yield

$$|u(z_o) - \Lambda_h u| =$$

$$(3.6)\quad = \frac{1}{4h^2} \Big| \int_0^1 \iint\limits_{\substack{|x-x_o|\leq h\\ |y-y_o|\leq h}} \{(x-x_o)u_x(x_o+t(x-x_o),y_o+t(y-y_o)) + (y-y_o)u_y(x_o+t(x-x_o),y_o+t(y-y_o))\}[dx,dy]dt\Big|$$

$$\leq \|u\|_{W_2^1}$$

for $u \in W_2^1 \cap C^o$. For $\Lambda_h u$ we find with the Schwarz inequality, (3.1) and (3.3) the estimate

[2] *For points near* ∂G *we use extensions of* u,v,V *in the desired spaces.*

$$|\Lambda_h u| \le |\Lambda_h(u-U)| + |\Lambda_h U|$$

$$(3.7) \qquad < \frac{1}{2h} \|u-U\|_{L^2} + \|U\|_{C^o}$$

$$\le c_1 \|u\|_{W_2^1} + M\, |\log h|^{\frac{1}{2}} \|u\|_{W_2^1} .$$

Hence the triangle inequality and $u = V-v$ yields with (3.1) the desired estimate:

$$|V(z_o) - v(z_o)| \le c_2 |\log h|^{\frac{1}{2}} \|V-v\|_{W_2^1} \le c\, |\log h|^{\frac{1}{2}} h^{s-1} \|v\|_{W_2^s} .$$

In the following, these finite element functions are used as to be trial functions in a least squares method for solving linear problems (1.7) with $g \equiv o$ and the Newton equations (1.11) approximatively. According to [27, chap. 8] this method reads as follows: Let $\tilde{\mathcal{H}}$ denote the linear space of elements

$$(3.8) \qquad \tilde{\chi} = \sum_{j=o}^{n-1} e^j (\chi_{j1} + i\chi_{j2}) = \sum_{j=o}^{n-1} e^j \tilde{\chi}_j \quad \text{with} \quad \chi_{jk} \in \tilde{H}$$

over the field of reals. Clearly, the approximation property (C) and stability property (S) as well as Lemma 3.1 and Lemma 3.2 remain valid with hypercomplex valued functions v, V and $\tilde{\mathcal{H}}$ instead of $\tilde{H}$. The least squares method for (1.7) with $g \equiv o$ yields the following Gauss equations:

Find $\tilde{w} \in \tilde{\mathcal{H}}$ *such that*

$$
\begin{aligned}
[\tilde{w},\tilde{\chi}] := & \iint_G \{D\tilde{w}, D\tilde{\chi}\}[dx,dy] + h^{-1} \oint_{\partial G} \{R\tilde{w},R\tilde{\chi}\}\, ds \\
& + \sum_{j=o}^{n-1} \Big(\sum_{\ell=1}^{m_j} \mathrm{Re}(\tilde{w}_j(z_\ell)\cdot\overline{\tilde{\chi}_j(z_\ell)}) + \oint_{\partial G} \sigma \mathrm{Im}\, p_{jj} w_j ds \oint_{\partial G} \sigma \mathrm{Im}\, p_{jj}\tilde{\chi}_j ds\Big) \\
(3.9) \qquad = & \iint_G \{f,D\tilde{\chi}\}[dx,dy] + h^{-1} \oint_{\partial G} \{\psi,R\tilde{\chi}\}\, ds \\
& + \sum_{j=o}^{n-1} \Big(\sum_{\ell=1}^{m_j} \mathrm{Re}\,(\gamma_{j\ell}\, \overline{\tilde{\chi}_j(z_\ell)}) + \kappa_j \oint_{\partial G} \sigma \mathrm{Im}\, p_{jj}\, \tilde{\chi}_j\, ds \Big)
\end{aligned}
$$

is satisfied for all test functions $\tilde{\chi} \in \tilde{\mathcal{H}}$ where $\{\ ,\ \}$ is defined as the scalar product for two hypercomplex a,b by

$$(3.10) \qquad \{\sum_{j=o}^{n-1} e^j a_j, \sum_{k=o}^{n-1} e^k b_k\} := \sum_{j=o}^{n-1} \mathrm{Re}\, a_j\, \overline{b}_j\ .$$

Since the linear problem (1.7) with $g \equiv o$ has a unique solution w due to Theorem 1.1, the equation (3.9) has also a unique solution $\tilde{w} \in \tilde{\mathcal{H}}$ to each $h > o$. Moreover, the bilinear form $[\ ,\ ]$ defined by (3.9) defines a scalar product and (3.9) is also equivalent to

$$
(3.11) \qquad
\begin{aligned}
& [w-\tilde{w},\tilde{\chi}] = o \quad \text{for all } \tilde{\chi} \in \tilde{\mathcal{H}} \quad \text{and} \\
& [w-\tilde{w},w-\tilde{w}] = \min_{\tilde{\chi}\in\tilde{\mathcal{H}}} [w-\tilde{\chi},w-\tilde{\chi}]
\end{aligned}
$$

where w denotes the solution to the original problem (1.7) with $g \equiv o$. Hence, the mapping from w onto $\tilde{w}$ is well defined and is called the *Gauss projection* $\tilde{G}$:

$$(3.12) \quad \underset{\sim}{G} : w \to \tilde{w} \quad \text{with} \quad \underset{\sim}{G}\,\tilde{\chi} = \tilde{\chi} \quad \text{for all} \quad \tilde{\chi} \in \tilde{\mathcal{H}}.$$

The convergence $\tilde{w} \to w$ for $h \to o$ in [27, Theorem 8.4.1] of optimal order was obtained under the assumption that the linear functionals involved are *continuous* on L^2. Since this is not the case for the point conditions in (1.7) and (3.9) we shall extend the results of [27] to our case following essentially the presentation in [27, § 8.4].

For the bilinear form in (3.9) we find

Lemma 3.3: There exists a constant c *independent of* h *and* v *such that for all* $1 \le s \le m+1$ *and the hypercomplex valued* $v \in W_2^s \quad C^o$ *the estimate*

$$(3.13) \quad \min_{\tilde{\chi} \in \tilde{\mathcal{H}}} [\tilde{\chi}-v,\ \tilde{\chi}-v]^{\frac{1}{2}} \le c\,|\log h|^{\frac{1}{2}}\, h^{s-1}\, \|v\|_{W_2^s}$$

holds.

Proof: Proceeding in exactly the same way as in the proof of [27, Lemma 8.4.2] we find the estimate

$$[\tilde{\chi}-v,\ \tilde{\chi}-v] \le c_1 \|\tilde{\chi}-v\|_{W_2^1} \{\|\tilde{\chi}-v\|_{W_2^1} + h^{-1} \|\tilde{\chi}-v\|_{L^2}\}$$
$$+ \sum_{j=o}^{n-1} \sum_{\ell=1}^{m_j} |\tilde{\chi}_j(z_\ell) - v_j(z_\ell)|^2 .$$

Choosing $\tilde{\chi} = V$ and using (3.1) and (3.4) we find (3.13).

Next we execute Nitsche's trick by using a-priori estimates for the error

$$(3.14) \qquad E := w - \tilde{w} = w - \underset{\sim}{G}w$$

in the topological dual spaces W_2^{-p} to W_2^p associated with each other by the L^2 duality defined by the scalar product

$$(3.15) \qquad (v,g)_{L^2} := \iint_G \{v,g\}\,[dx,dy].$$

The norm is defined by

$$\|g\|_{W_2^{-p}} = \sup_{\|h\|_{W_2^p} \le 1} (g,h)_{L^2} \,.$$

Correspondingly we introduce on the boundary the scalar product

$$(3.16) \qquad \langle\psi,\phi\rangle_{L^2} := \oint_{\partial G} \{\psi,\phi\}\, ds$$

and we also have $W_2^{-q}(\partial G)$ spaces (see e. g. in [2], [3]).

Lemma 3.4: For $1 \le s \le m+1$, $o \le p \le m$, $-\frac{1}{2} \le q \le m$ *and* $w \in W_2^s \cap C^o$ *we have the following estimates:*

$$(3.17) \qquad [E,E]^{\frac{1}{2}} \le c\,|\log h|^{\frac{1}{2}}\, h^{s-1}\, \|w\|_{W_2^s}\,,$$

$$(3.18) \qquad \|DE\|_{W_2^{-p}} \le c|\log h|^{\sigma} h^{s+p-1}\, \|w\|_{W_2^s} \quad \text{where } \sigma = \tfrac{1}{2} \text{ for } p=o \text{ and } \sigma = 1 \text{ for } p>o,$$

(3.19) $$\|RE\|_{W_2^{-q-\frac{1}{2}}(\partial G)} \leq c|\log h|^{\sigma} h^{s+q} \|w\|_{W_2^s}$$

where $\sigma = \frac{1}{2}$ for $q = -\frac{1}{2}$ and $\sigma = 1$ for $q > -\frac{1}{2}$,

(3.20) $$\left|\oint_{\partial G} \sigma \operatorname{Im} p_{jj} E\, ds\right| \leq c|\log h|^{\sigma} h^{s+p-1} \|w\|_{W_2^s},$$

with σ as in (3.18),

(3.21) $$|E_j(z_\ell)| \leq c|\log h|^{\sigma} h^{s+p-1} \|w\|_{W_2^s}$$

with σ as in (3.18)

Proof: From (3.11), (3.14) and (3.13) we see immediately (3.17). For $p = o$ and $q = -\frac{1}{2}$ the inequalities (3.18) - (3.21) follow from (3.17) and the definition of [,]. Now let $p > o$ and $v \in W_2^{p+1}$ be an arbitrary solution of

(3.22) $$Dv = g,\ Rv = \phi,\ v_j(z_\ell) = \alpha_{j\ell},\ \oint_{\partial G} \sigma \operatorname{Im} p_{jj} v_j ds = \nu_j .$$

Since (3.22) is uniquely solvable, v traces W_2^{p+1} if the right hand sides trace $W_2^p \times W_2^{p+\frac{1}{2}}(\partial G) \times \mathbb{C}^{\Sigma m_j} \times \mathbb{R}^n$. Moreover, we have the a-priori estimate (2.3). Using (3.11), (3.14), and (3.1) we find with (3.13)

$$[E,v] = [E,v-\tilde{\chi}] \quad \text{for every } \tilde{\chi} \in \tilde{H}$$

$$\le [E,E]^{\frac{1}{2}} \min_{\tilde{\chi}\in\tilde{H}} [E,v-\tilde{\chi}]$$

$$\le [E,E]^{\frac{1}{2}} |\log h|^{\frac{1}{2}} h^p \|v\|_{W_2^{p+1}} .$$

Using (3.17) and the a-priori estimate (2.3) we find from (3.9),

$$(DE,g)_{L^2} + h^{-1} \langle Re,\Phi\rangle$$

$$(3.23) \quad + \sum_{j=o}^{n-1} \Big(\sum_{\ell=1}^{m_j} Re(E_j(z_\ell)\bar{\alpha}_{j\ell}) + \int_{\partial G} \sigma Im\, p_{jj}E_j ds\cdot\nu_j\Big) = [E,v]$$

$$\le c|\log h| h^{s+p-1} \|w\|_{W_2^s} \{\|g\|_{W_2^p} + \|\Phi\|_{W_2^{p+\frac{1}{2}}(\partial G)}$$

$$+ \sum_{j=o}^{n-1} \Big(\sum_{\ell=1}^{m_j} |\alpha_{j\ell}| + |\nu_j|\Big)\}.$$

Choosing $\phi = o$, $\alpha_{j\ell} = o$, $\nu_j = o$ and g tracing W_2^p one finds (3.18). Using appropriate other choices we obtain also (3.19) - (3.21). Interpolation between the two extreme values $p = o$ and $p = m$ as well as $q = -\frac{1}{2}$ and $q = m$ yields uniform constants.

Now we can prove the announced convergence results for the linear problems.

Theorem 3.5: For $1 \leq t \leq s$, $t \leq m$, $1 \leq s \leq m+1$, $w \in C^{\alpha} \cap W_2^s$, $\alpha > 0$, *there exist constants* c *which are independent of* h *and the solution* w *of* (1.7) *with* $g \equiv 0$ *such that the asymptotic error estimate*

$$(3.24) \quad \|\tilde{w}-w\|_{C^0} + \|\tilde{w}-w\|_{W_2^t} \leq c|\log h| h^{s-t} \|w\|_{W_2^s}$$

holds. The Gauss projections satisfy

$$(3.25) \quad \|\underset{\sim}{G}w\|_{W_2^1} \leq c(|\log h| h^{s-1} \|w\|_{W_2^s} + \|w\|_{W_2^1})$$

and

$$(3.26) \quad \|\underset{\sim}{G}w\|_{C^0} \leq \|w\|_{C^0} + c|\log h| h^{s-1} \|w\|_{W_2^s} .$$

Proof: From [27, § 1.2, § 5.2] we know that the solutions of the problem

$$(3.27) \quad Dh = 0, \; Rh = 0$$

form a $n + \sum_{j=0}^{n-1} m_j = N$-dimensional space of smooth functions $h \in C^{1+\alpha} \cap W_2^{m+1}$. Let h_ℓ, $\ell = 1,\dots,N$ be a corresponding basis orthogonalized with respect to

$$(3.28) \quad (h_\ell, h_m)_{L^2} = \delta_{\ell,m} , \quad \ell,m = 1,\dots,N.$$

Then the linear functionals defined by

(3.29) $\Lambda_\ell v := (v, h_\ell)_{L^2}$

are continuous on L^2 satisfying (A.6). Now let us write E in the form

(3.30) $E = E_1 + E_2 + E_3$

where the E_j are defined as to be the solutions of

(3.31) $DE_1 = DE, \; RE_1 = o, \; \Lambda_\ell E_1 = o, \quad \ell = 1,\dots,N;$

(3.32) $DE_2 = o, \quad RE_2 = RE, \; \Lambda_\ell E_2 = o, \; \ell = 1,\dots,N;$

(3.33) $DE_3 = o, \quad RE_3 = o,$

$$E_{3j}(z_k) = E_j(z_k) - E_{1j}(z_k) - E_{2j}(z_k),$$

$$\int_{\partial G} \sigma \mathrm{Im}\, p_{jj}\, E_{3j}\, ds = \oint_{\partial G} \sigma \mathrm{Im}\, p_{jj}(E_j - E_{1j} - E_{2j})\, ds.$$

Since $E = w - \mathcal{G}w \in C_1^\alpha \cap W_2^1$ and $E_3 \in C^{1+\alpha} \cap W_2^{m+1}$ it is easily seen that $E_2 \in C^\alpha \cap W_2^1$ and $E_1 \in C_1^\alpha \cap W_2^1$.
Using Dikanski 's estimate (2.6) for E_1 with (3.18) we have

(3.34) $\|E_1\|_{W_2^1} \le c|\log h|^{\frac{1}{2}} h^{s-1} \|w\|_{W_2^s}$.

Similarly (2.6) yields for E_2 with (3.19)

(3.35) $\|E_2\|_{L^2} \le c|\log h|^{\frac{1}{2}} h^{s} \|w\|_{W_2^s}$.

Form (2.1) and (2.3) we find for E_3:

$$\|E_3\|_{W_2^{m+1}}+\|E_3\|_{C^{1+\alpha}} \le k_1\{\sum_{j=o}^{n-1} (|\oint_{\partial G}\sigma \mathrm{Im}\ p_{jj}(E_j-E_{1j}-E_{2j})ds| + \sum_{\ell=1}^{m_j} |E_{3j}\ (z_\ell)|\} \tag{3.36}$$

Clearly, using (3.20), (3.21) and (3.34), (3.35), this estimate becomes

$$\|E_3\|_{W_2^{m+1}} + \|E_3\|_{C^{1+\alpha}} \le c|\log h|^{\frac{1}{2}}\ h^{s-1}\ \|w\|_{W_2^s} + c' \sum_{j=o}^{n-1} \sum_{\ell=o}^{m_j} (|E_{1j}(z_\ell)|+|E_{2j}(z_\ell)|). \tag{3.37}$$

Hence, we need estimates for the values $E_{1j}(z_\ell)$ and $E_{2j}(z_\ell)$.

To this end we use (3.4) and (3.1) for E_1; (3.34) implies

$$|E_{1j}(z_\ell)| \le \|E_1\|_{C^o} \le c|\log h|h^{s-1}\ \|w\|_{W_2^s}\ . \tag{3.38}$$

Since for E_2 we only have an L^2 estimate (3.35) yet we shall approximate E and E_1, E_2, E_3 by suitable finite elements. Here we need the orthogonal projection P of L^2 with (3.15) onto $\mathcal{H}$: P : $L^2 \to \mathcal{H}$. Then in (3.1) and, consequently in (3.4) V can be replaced by Pv:

$$\|Pw-w\|_{W_2^k} \le c'_{rk}\ h^{r-k}\ \|w\|_{W_2^r} \tag{3.39}$$

and

$$(3.40)\quad |Pw(z_o) - w(z_o)| \le c'|\log h|^{\frac{1}{2}} h^{s-1} \|w\|_{W_2^s}$$

(see [9], [18]). With linearity of P and triangle inequality we have

$$\|E_2\|_{W_2^1} \le \|PE-E\|_{W_2^1} - \|PE_1-E_1\|_{W_2^1} + \|PE_3-E_3\|_{W_2^1} + \|PE_2\|_{W_2^1}$$

With (3.12), (3.14), (3.39), (3.1), (3.2) we find

$$\|E_2\|_{W_2^1} \le \|Pw-w\|_{W_2^1} + c\|E_1\|_{W_2^1} + c'h^m\|E_3\|_{W_2^{m+1}} + Mh^{-1}\|E_2\|_{L^2}$$

Inserting (3.34), (3.35), (3.36), (3.37) and (3.39) we find

$$(3.41)\quad \|E_2\|_{W_2^1} \le c|\log h|^{\frac{1}{2}} h^{s-1} \|w\|_{W_2^s} + c'h \sum_{j=o}^{n-1} \sum_{\ell=1}^{m_j} |E_{3j}(z_\ell)|.$$

and

$$(3.42)\quad |E_{2j}(z\ )| \le \|E_2\|_{C^o} \le c''|\log h| h^{s-1} \|w\|_{W_2^s} + $$
$$+ c''|\log h|^{\frac{1}{2}} \cdot h \cdot \sum_{j=o}^{n-1} \sum_{\ell=1}^{m_j} |E_{3j}(z_\ell)|.$$

(3.37) and (3.42) imply an inequality of the form

$$(3.43)\quad |E_{3j}(z_\ell)| \le \|E_3\|_{C^o} \le c|\log h| h^{s-1} \|w\|_{W_2^s} +$$
$$+ c'h|\log h|^{\frac{1}{2}} \sum_{j=o}^{n-1} \sum_{\ell=1}^{m_j} |E_{3j}(z_\ell)|.$$

choosing $h_o > o$ suitably small, (3.43) yields

$$(3.44)\quad |E_{3j}(z_\ell)| \le c|\log h| h^{s-1} \|w\|_{W_2^s}$$

and, consequently with (3.37)

$$(3.45)\quad \|E_3\|_{W_2^{m+1}} + \|E_3\|_{C^{1+\alpha}} \le c|\log h| h^{s-1} \|w\|_{W_2^s} .$$

Using (3.44) in (3.41) and (3.42) we also find

$$(3.46)\quad \|E_2\|_{W_2^1} \le c|\log h|^{\frac{1}{2}} h^{s-1} \|w\|_{W_2^s}$$

and

$$(3.47)\quad \|E_2\|_{C^o} \le c|\log h| h^{s-1} \|w\|_{W_2^s} .$$

Collecting (3.34), (3.38), (3.45) - (3.47) one obtains (3.24) with $t = 1$ and (3,25), (3,26) from $\tilde{w} = \underset{\sim}{G}w$ with the triangle inequality.

Now (3.24) follows by simple use of (3.1) and (3.2) from (3.25)

$$\begin{aligned}\|\underset{\sim}{G}w\|_{W_2^t} &\le \|V-w\|_{W_2^t} + \|\underset{\sim}{G}(V-w)\|_{W_2^t} \\ &\le c\, h^{s-t} \|w\|_{W_2^s} + Mh^{1-t} \|\underset{\sim}{G}(V-w)\|_{W_2^1} \\ &\le c' h^{s-t} |\log h|\, \|w\|_{W_2^s}.\end{aligned}$$

Remark 3.6: The point conditions are implemented by the corresponding least squares expressions in (3.9). It is not very difficult to show the same estimates as in Theorem 3.5 for approximate solutions satisfying the *point conditions itself*:

$$(3.48) \quad \tilde{w}_j(z_\ell) = \gamma_{j\ell}; \quad \ell = 1,\ldots,m_j, \; j = o,\ldots,n-1$$

§ 4 THE IMBEDDING AND NEWTON'S METHOD WITH FINITE ELEMENTS

For the nonlinear problems (1.7) we shall now define a suitable finite element approximation. To this end we shall assume throughout this section that (A.3) is satisfied and instead of (1.7) we shall solve the problems (1.10). The proposed version of imbedding and Newton approximation is executed as follows: For fixed h, for fixed $t \in [o,1]$ (which will be specified by $t = t_j$, $o = t_o < t_1 < \ldots < t_N$) and choosen $\tilde{w}^{(o)} \in \tilde{\mathcal{H}}$ compute a sequence of finite element functions $\tilde{w}^{(1)}, \tilde{w}^{(2)},\ldots,\tilde{w}^{(m+1)},\ldots$ by solving for each $m = o,1\ldots$ the approximate *linear* problems

$$
\begin{aligned}
&(D\tilde{w}^{(m+1)} - t\sum_{\ell=o}^{n-1}(\sum_{k=o}^{\ell} e^{\ell}(\frac{\partial g_\ell}{\partial w_k}(z,\tilde{w}^{(m)})(\tilde{w}_k^{(m+1)}-\tilde{w}_k^{(m)})\\
(4.1)\quad &+\frac{\partial \tilde{g}_\ell}{\partial \overline{w}_k}(z,\tilde{w}^{(m)})(\overline{\tilde{w}}_k^{(m+1)}-\overline{\tilde{w}}_k^{(m)})) - t\tilde{g}(z,\tilde{w}^{(m)})-f,D\tilde{\chi})_{L^2}\\
&+h^{-1}<R\tilde{w}^{(m+1)} - \psi, R\tilde{\chi}>_{L^2(\partial G)}\\
&+\sum_{j=o}^{n-1}\{\sum_{\ell=1}^{m_j}(\tilde{w}_j^{(m+1)}(z_\ell)-\gamma_{j\ell})\tilde{\chi}_j(z_\ell))+(\oint_{\partial G}\sigma \mathrm{Im}(p_{jj}\tilde{w}_j^{(m+1)})ds\\
&\cdot \oint_{\partial G}\sigma \mathrm{Im}(p_{jj}\tilde{\chi}_j)ds)\} = o.
\end{aligned}
$$

Clearly, each of these problems is of the form (3.9), a system of (dim $\tilde{H}$) linear equations for the coefficients of $\tilde{w}^{(m+1)}$ with a sparse matrix of coefficients if we use an appropriate basis of $\tilde{H}$. Each of these problems is uniquely solvable. With the choice of a suitable sequence $t_o = o < t_1 < \ldots < t_N = 1$ and with

$$(4.2)\qquad \tilde{w}^{(o)}(z;t_j) := \lim_{m\to\infty} \tilde{w}^{(m)}(z,t_{j-1})$$

we have a discretized version of the Newton imbedding (1.11). Hence, it is a natural question to ask for the convergence of the method in connection with a possible choice of finitely many $t_1,\ldots,t_{N-1}$; and also to ask for asymptotic estimates for $w-\tilde{w}(z;1)$. We shall prove that, indeed, for fixed $h > o$, the method (4.1) converges for $m \to \infty$ and all $t_j \in [o,1]$ and for all $h \le h_o$ uniformly where the $t_j < t_{j+1}$ can be found appropriately. Of course, in practice the *limit* in (4.2) will be replaced by some *suitable iterate* ($m < \infty$).

For the further investigation the equations (4.2) will be replaced *formally* by the following system of coupled problems:

$$(4.3)\quad \tilde{w}^{(m+1)} = t\underset{\sim}{G}v^{(m+1)} + t\underset{\sim}{G}f^{(m)} + \underset{\sim}{G}w(.,o),$$

$$Dv^{(m+1)} = \sum_{\ell=o}^{n-1} e^{\ell} \sum_{k=o}^{\ell} \left(\frac{\partial\tilde{g}_{\ell}}{\partial w_k}(z,\tilde{w}^{(m)})\tilde{w}_k^{(m+1)} + \frac{\partial\tilde{g}_{\ell}}{\partial\overline{w}_k}(z,\tilde{w}^{(m)})\overline{\tilde{w}}_k^{(m+1)}\right)$$

$$(4.4)\quad \begin{aligned} Rv^{(m+1)} = o,\ v_j^{(m+1)}(z_{\ell}) &= o, \quad \ell = 1,\dots,m_j, \\ \oint_{\partial G} \sigma \mathrm{Im}\, p_{jj} v_j^{(m+1)} &= o, \quad j = o,\dots,n-1, \end{aligned}$$

$$(4.5)\quad \begin{aligned} Df^{(m)} = &- \sum_{\ell=o}^{n-1} e^{\ell} \sum_{k=o}^{\ell} \left(\frac{\partial\tilde{g}_{\ell}}{\partial w_k}(z,\tilde{w}^{(m)})\tilde{w}_k^{(m)} + \right. \\ &+ \left.\frac{\partial\tilde{g}_{\ell}}{\partial\overline{w}_k}(z,\tilde{w}^{(m)})\overline{\tilde{w}}_k^{(m)}\right) + \tilde{g}(z,\tilde{w}^{(m)}) \end{aligned}$$

$$Rf^{(m)} = o,\ f_j^{(m)}(z_{\ell}) = o \quad \ell = 1,\dots,m_j,$$

$$\oint_{\partial G} \sigma \mathrm{Im}\, p_{jj} f_j^{(m)}\, ds = o \quad j = o,\dots,n-1.$$

Here $w(.,o) \in C^{1+\alpha} \cap W_2^s$ denotes the solution of (1.10) for $t = o$. In order to find estimates for $\tilde{w}^{(m+1)}$ and $(\tilde{w}^{(m+1)} - \tilde{w}^{(m)})$ let us investigate problems like (4.3) - (4.5), i. e. problems of the form

$$(4.6)\quad \tilde{w} = t\underset{\sim}{G}v + u,$$

$$(4.7) \quad Dv = \sum_{\ell=o}^{n-1} e^{\ell} \sum_{k=o}^{\ell} (\tilde{g}_{\ell k} \tilde{w}_k + \tilde{\tilde{g}}_{\ell k} \overline{\tilde{w}_k}) + f ,$$

$$Rv = o, \; v_j(z_\ell) = o, \; \ell = 1,\ldots,m_j, \; \oint_{\partial G} \sigma \mathrm{Im}\, p_{jj} v_j ds = o,$$

$$j = o,\ldots,n-1.$$

Let u be smooth enough, $f \in L^p$, $\tilde{g}_{\ell k}$, $\tilde{\tilde{g}}_{\ell k} \in L^\infty$ and let

$$(4.8) \quad \|\tilde{g}_{\ell k}\|_{L^\infty} , \quad \|\tilde{\tilde{g}}_{\ell k}\|_{L^\infty} \le K_1 .$$

Then we have

Lemma 4.1: There exists a constant K_2 *which is independent of* h,f,u, t $\in$ [o,1] *and the coefficients* $\tilde{g}_{\ell k}$, $\tilde{\tilde{g}}_{\ell k}$ *satisfying* (4.8) *such that the a-priori estimates*

$$(4.9) \quad \|\tilde{w}\| \le K_2 \; (\|f\|_{L^p} + \|u\|)$$

hold with $\|.\|$ *the maximum norm in* C^o, *the* W_2^1 *norm, the* $W_2^{1+\varepsilon}$ *norm, the* C^γ *norm or the* W_p^1 *norm.* (4.9) *holds uniformly in either case,* $\varepsilon > o$, $\gamma = \frac{p-2}{2} > \frac{1+\varepsilon}{2}$.

Proof: Before presenting the details of the proof let us recall some imbedding properties concerning different function spaces. Choose β with $\gamma > \beta > \frac{1+\varepsilon}{2}$.

Then by using the Besov space $B_2^{\frac{1}{2}(1+\varepsilon),2}$ [1, § 7.71, § 7.72] we have an estimate

$$\| \phi \|_{W_2^{1+\varepsilon}} \leq c \| \phi \|_{B_2^{\frac{1}{2}(1+\varepsilon),2}}$$

(4.10) $$= x \, \{ \iint_G \iint_{G_z} \frac{|\phi(z)-\phi(\frac{1}{2}(z-\zeta)-\phi(\zeta)|^2}{|z-\zeta|^{3+\varepsilon}} [dx,dy][d\xi,d\eta] \}^{\frac{1}{2}}$$

$$\leq c' \| \phi \|_{C^\beta} \, .$$

Hence, the following imbedding properties are valid:

(4.11) $C^\beta \hookrightarrow W_2^{1+\varepsilon}$ *is continuous for* $\beta > \frac{1+\varepsilon}{2}$,

(4.12) $C^\gamma \hookrightarrow C^\beta$ *is compact for* $o < \beta < \gamma$ [29, p. 260]

(4.13) $W_2^{1+\varepsilon} \hookrightarrow C^o$ *is continuous for* $\varepsilon > o$ [1, 7.63-7.66].

(4.14) $\mathfrak{W}^1_{p,\gamma} \hookrightarrow C^\gamma$ is continuous for $\gamma = \frac{p-2}{p}$

where $\mathfrak{W}^1_{p,\gamma} := \{ w | w_{\bar{z}} \in L^p \wedge Rw \in C^\gamma(\partial G) \cap W_p^{1-\frac{1}{p}}(\partial G)$

$$\|w\|_{\mathfrak{W}^1_{p,\gamma}} := \|w\|_{W_p^1} + \|Rw\|_{C^\gamma(\partial G)} + \|Rw\|_{W_p^{1-\frac{1}{p}}(\partial G)} \}$$

(see [23, section 6], [12] , [14])

Now (4.9) will be proved for the C^o-norm by contradiction. For the remaining cases, the proof can be executed correspondingly.

Let us assume that there exist sequences

$$\tilde{w}^{(j)}, h^{(j)} \to h^{(\infty)} \geq o,\ t_j \to t_o \in [o,1],\ f^{(j)},\ u^{(j)}$$

(4.15) $\tilde{g}_{\ell kj}, \tilde{\tilde{g}}_{\ell kj}$ such that

$$(\|f^{(j)}\|_{L^p} + \|u^{(j)}\|_{C^o})\ (\|\tilde{w}^{(j)}\|_{C^o})^{-1} \to o.$$

Now we devide (4.6) by $\|\tilde{w}^{(j)}\|_{C^o}$ introducing

$$(4.16)\quad w^{(j)} := \tilde{w}^{(j)}(\|\tilde{w}^{(j)}\|_{C^o})^{-1},\ \text{i. e.}\ \|w^{(j)}\|_{C^o} = 1,$$

$$Dv^{(j)} = \sum_{\ell=o}^{n-1} e^{\ell} \sum_{k=o}^{\ell} (\tilde{g}_{kj}\ w_k^{(j)} + \tilde{\tilde{g}}_{kj}\ \overline{w}_k^{(j)}) + f^{(j)}(\|\tilde{w}^{(j)}\|_{C^o})^{-1}$$

and homogeneous boundary and point conditions as in (4.4). Hence, we have a *uniform bound*

$$(4.17)\quad \|Dv^{(j)}\|_{L^p} \leq K_3$$

from (4.8). The a-priori estimate (2.2) implies the uniform bound

$$(4.18)\quad \|v^{(j)}\|_{C^{\gamma}} \leq c\,\|v^{(j)}\|_{W_p^1} \leq K_4.$$

Therefore, we can find a subsequence of $v^{(j)}$ with

$$(4.19)\quad v^{(j)} \to v^{(\infty)} \quad \text{in } C^{\beta} \text{ and in } W_2^{1+\varepsilon}$$

due to (4.12) and (4.11). With this sequence we find from

(4.6) and the estimate (3.24) together with (4.13) and (4.16)

$$(4.20)\quad w^{(j)} = t_j \underset{\sim}{G} v^{(j)} + u^{(j)} (\|\tilde{w}^{(j)}\|_{C^o})^{-1} \to w^{(\infty)} = t_o \underset{\sim}{G}_{(h^{(\infty)})} v^{(\infty)}$$

$$\text{with } \|w^{(\infty)}\|_{C^o} = 1 .$$

The convergence holds in C^o whereas the limit is in $W_2^{1+\varepsilon'}$, $o < \varepsilon' < \varepsilon$. Since we have (4.8) we can pass to a new subsequence such that

$$(4.21)\quad \tilde{g}_{\ell kj} \rightharpoonup \tilde{g}_{\ell k\infty},\ \tilde{\tilde{g}}_{\ell kj} \rightharpoonup \tilde{\tilde{g}}_{\ell k\infty} \quad \text{and } \|g_{\ell k\infty}\|_{L^\infty},$$

$$\|g_{\ell k\infty}\|_{L^\infty} \leq K_1 ,$$

converge *weakly* in L^p for any p with $2 < p < \infty$.
Hence, besides

$$(4.22)\quad w^{(\infty)} = t_o \underset{\sim}{G}_{(h^{(\infty)})} v^{(\infty)}$$

we find easily that $v^{(\infty)} \varepsilon C^\beta \cap W_2^{1+\varepsilon}$ is a weak solution of the linear system

$$(4.23)\quad Dv^{(\infty)} = \sum_{\ell=o}^{n-1} e^\ell \sum_{k=o}^{\ell} (\tilde{g}_{\ell k\infty} w_k^{(\infty)} + \tilde{\tilde{g}}_{\ell k\infty} \overline{w}_k^{(\infty)})$$

satisfying the homogeneous boundary and point conditions as in (4.4). If $h^{(\infty)} = o$ then $\underset{\sim}{G}$ is the identity and $v^{(\infty)} = w^{(\infty)}$ satisfies (4.23) and also the homogeneous conditions (4.4). Theorem 1.1 implies $w^{(\infty)} \equiv o$, a contradiction to (4.20). For $h^{(\infty)} > o$ we find from (4.22) and (4.23) the equation

$$(4.24)\quad Dv^{(\infty)} = t_o \sum_{\ell=o}^{n-1} e^\ell \sum_{k=o}^{\ell} (\tilde{g}_{\ell k\infty} v_k^{(\infty)} + \tilde{\tilde{g}}_{\ell k\infty} \overline{v}_k^{(\infty)})$$

$$\sum_{\ell=o}^{n-1} e^\ell \sum_{k=o}^{\ell} (\tilde{g}_{\ell k}((\underset{\sim}{G}_{(h^{(\infty)})} - I)v^{(\infty)})_k +$$

$$\tilde{\tilde{g}}_{\ell k\infty} \overline{(\underset{\sim}{G}_{(h^{(\infty)})} - I)v^{(\infty)})_k}) .$$

Hence, the a-priori estimate (2.2) yields with (4.21)

$$\|v^{(\infty)}\|_{C^\gamma} \le K_\gamma K_1 \| (\underset{\sim}{G}_{(h^{(\infty)})} - I)v^{(\infty)}\|_{C^o}$$

where $K_\gamma K_1$ is independent of $h^{(\infty)} \le h_o$ and $v^{(\infty)}$. Using (4.13) and (3.24) we find

$$\|v^{(\infty)}\|_{C^\gamma} \le c\ h_o^\varepsilon\ |\log h_o|\ \|v^{(\infty)}\|_{W_2^{1+\varepsilon}}$$

and with (4.11)

$$(4.25)\quad \|v^{(\infty)}\|_{W_2^{1+\varepsilon}} \le c\|v^{(\infty)}\|_{C^\gamma} \le c' h_o^\varepsilon |\log h_o|\ \|v^{(\infty)}\|_{W_2^{1+\varepsilon}}$$

$$\le \frac{1}{2} \|v^{(\infty)}\|_{W_2^{1+\varepsilon}}$$

if we choose $h_o > o$ and all $h \le h_o$ small enough throughout the paper.

Thus $v^{(\infty)} \equiv o$ and $w^{(\infty)} \equiv o$ in contradiction to (4.20).

As a simple consequence of Lemma 4.1 we have

Lemma 4.2: For any $\tilde{w}^{(m)}$ *the solution* $\tilde{w}^{(m+1)}$ *of the Gauss equations* (4.1) *satisfies an a-priori estimate*

$$(4.26) \qquad \|\tilde{w}^{(m+1)}\|_{C^o} \le K$$

where K *is independent of* $\tilde{w}^{(m)}$, *of* h *and of* $t \in [o,1]$ *but depends on the right hand sides* $f \in C^\alpha \cap W_2^\varepsilon$ *and* $\psi_j \in C^{1+\alpha}(\partial G)$.

Proof: According to (4.3), (4.5) - (4.7) and Lemma 4.1 we choose

$$(4.27) \qquad u := t\underset{\sim}{G}f^{(m)} + \underset{\sim}{G}w(o).$$

Since $f^{(m)}$ is defined by (4.5), we find from the definition of $\tilde{g}$ in (1.8) that the whole right hand side in (4.5) is *continuous and equibounded* by

$$(4.28) \qquad \|Df^{(m)}\|_{C^o} \le K_7.$$

Again, with the a-priori estimate (2.2) we find using (4.11), (4.12) that the $f^{(m)}$ are *uniformly* bounded,

$$(4.29) \qquad \|f^{(m)}\|_{W_2^{1+\varepsilon}} \le c\|f^{(m)}\|_{C^\gamma} \le K_8.$$

Due to Theorem 2.1 we have that $\|w(o)\|_{C^{1+\alpha}} + \|w(o)\|_{W_2^{1+\varepsilon}}$ exists. Hence, with (3.26) there holds

$$(4.30)\qquad \|u\|_{C^o} \le c(K_8 + \|w(o)\|_{W_2^{1+\varepsilon}}) + \|w(o)\|_{C^{1+\alpha}} .$$

uniformly. Applying Lemma 4.1 we find the desired result.

Now we are in the position to prove one of our main results.

Theorem 4.3: There exists an $h_o > o$ *and a finite partition* $o = t_o < t_1 < \ldots < t_N = 1$ *such that the discrete Newton imbedding method* (4.1), (4.2) *converges in* C^o *for every* $h \le h_o$.

Proof: Using the Taylor formula as in (1.11), (1.12) we finally find from (4.3) that

$$(4.31)\qquad (\tilde{w}^{(m+1)} - \tilde{w}^{(m)}) = \underset{\sim}{G} v^{(m)}$$

where $v^{(m)}$ is defined by the solutions of

$$(4.32)\ Dv^{(m)} = \sum_{\ell=o}^{n-1} e^{\ell} \sum_{k=o}^{\ell} t_j \left(\frac{\partial \tilde{g}_\ell}{\partial w_k}(z,\tilde{w}^{(m)})(\tilde{w}_k^{(m+1)} - \tilde{w}_k^{(m)}) + \right.$$

$$\left. + \frac{\partial \tilde{g}_\ell}{\partial \overline{w}_k}(z,\tilde{w}^{(m)})(\overline{\tilde{w}}_k^{(m+1)} - \overline{\tilde{w}}_k^{(m)})\right) + f^{(m)}$$

and (4.4) with

$$(4.33)\quad f^{(m)} = t_j \sum_{\ell=o}^{n-1} e^{\ell} \sum_{p,k=o}^{\ell} \tilde{g}_{\ell kp11}(z,\tilde{w}^{(m)},\tilde{w}^{(m-1)}) \cdot$$

$$\cdot \{\tilde{w}_k^{(m)} - \tilde{w}_k^{(m-1)}\}\{\tilde{w}_p^{(m)} - \tilde{w}_p^{(m-1)}\}$$

$$+ 2\tilde{g}_{\ell kp12} \{\tilde{w}_k^{(m)} - \tilde{w}_k^{(m-1)}\}\{\overline{\tilde{w}}_p^{(m)} - \overline{\tilde{w}}_p^{(m-1)}\}$$

$$+ \tilde{g}_{\ell kp22} \{\overline{\tilde{w}}_k^{(m)} - \overline{\tilde{w}}_k^{(m-1)}\}\{\overline{\tilde{w}}_p^{(m)} - \overline{\tilde{w}}_p^{(m-1)}\} .$$

Using Lemma 4.1 with $\tilde{w} = \tilde{w}^{(m+1)} - \tilde{w}^{(m)}$ and $n = o$ we find

$$(4.34)\quad \|\tilde{w}^{(m+1)} - \tilde{w}^{(m)}\|_{C^o} \le K_2 \|f^{(m)}\|_{C^o} \le t_j K_9 \|\tilde{w}^{(m)} - \tilde{w}^{(m-1)}\|^2_{C^o}$$

uniformly, i. e. K_9 is independent of m, $h \le h_o$, t_j and the data in (1.11).

For the initial approximates we find

$$(4.35)\quad \tilde{w}^{(1)} - \tilde{w}^{(o)} = \tilde{G}v'$$

where v' is defined by (4.34) with $m = 1$ and

$$(4.36)\quad f^{(1)} = (t_j - t_{j-1})\tilde{g}(z,\tilde{w}^{(o)})$$

instead of (4.33). Again, by Lemma 4.1 we find from the equiboundedness of $\tilde{g}$ the estimate

$$(4.37) \quad \|\tilde{w}^{(1)} - \tilde{w}^{(o)}\|_{C^o} \leq (t_j - t_{j-1})K$$

with K independent of h, t_j, $w^{(o)}$. Hence, the only requirement is

$$(4.38) \quad t_j \, K_8 (t_j - t_{j-1}) K < 1$$

in order to assure the quadratic convergence in C^o.

As a consequence, the Newton imbedding method provides the construction of $\tilde{w}(t_N)$ as the limit for $m \to \infty$ at $t_N = 1$, the solution of

$$\iint_G \{D\tilde{w} - \tilde{g}(z,\tilde{w}) - c, D\tilde{\chi}\} \, [dx,dy]$$

$$h^{-1} \oint_{\partial G} \sum_{j=o}^{n-1} \Big(\sum_{k=o}^{j} \mathrm{Re}(p_{jk}\tilde{w}_k) - \psi_j \Big) \Big(\sum_{k=o}^{j} \mathrm{Re}(p_{jk}\tilde{\chi}_k) \Big) ds$$

$$+ \sum_{j=o}^{n-1} \Big(\sum_{\ell=1}^{m_j} \mathrm{Re}(\tilde{w}_j(z_\ell) \cdot \overline{\tilde{\chi}_j(z_\ell)}) + \Big(\oint_{\partial G} \sigma \mathrm{Im}(p_{jj} w_j) ds - \kappa_j \Big) \cdot$$

$$\cdot \oint_{\partial G} \sigma \mathrm{Im}(p_{jj}\tilde{\chi}_j) ds \Big) = o$$

for all $\tilde{\chi} \in H$.

For this finite element approximation we have the following asymptotic error estimate:

Theorem 4.4: For $1 \leq t \leq s \leq m+1$, $t \leq m$, $w \in C^{\alpha} \cap W_2^s$ *there exists a constant* c *which is independent of* h *and* w *such that the asymptotic estimates*

$$(4.40) \qquad \|\tilde{w}-w\|_{C^o} \leq c\,|\log h|\,h^{s-1}\,\|w\|_{W_2^s} \qquad \textit{and}$$

$$(4.41) \qquad \|\tilde{w}-w\|_{W_2^t} \leq c\,|\log h|\,h^{s-t}\,\|w\|_{W_2^s}$$

hold.

Proof: From (4.39) we can write

$$(4.42) \qquad \tilde{w} = \underset{\sim}{G}\tilde{w} = \underset{\sim}{G}v + \underset{\sim}{G}w$$

where v is defined by

$$(4.43) \qquad Dv = \tilde{g}(z,\tilde{w})-\tilde{g}(z,w) = \sum_{\ell=o}^{n-1} e^{\ell} \sum_{k=o}^{\ell} (g'_{k\ell}(\tilde{w}_k-w_k)+g''_{k\ell}(\overline{\tilde{w}_k}-\overline{w}_k))$$

satisfying the homogeneous conditions as in (4.4). Hence, from Lemma 4.1 we have for

$$\tilde{w} - w = \underset{\sim}{G}v + (\underset{\sim}{G}-I)w$$

with $f = o$ and $u = (\underset{\sim}{G}-I)w$ the estimate

$$(4.44) \qquad \|\tilde{w}-w\|_{C^o} \leq K\,\|(\underset{\sim}{G}-I)w\|_{C^o} \qquad \text{and}$$

$$(4.45) \qquad \|w-w\|_{W_2^1} \leq K\,\|(\underset{\sim}{G}-I)w\|_{W_2^1} \quad .$$

Now (3.24) with $t = 1$ in (4.44) yields (4.40) and with $t \geq 1$ in (4.45) yields (4.41).

Remark 4.5: Again, Theorem 4.4 remains valid if we use in (4.1) the point conditions

$$\tilde{w}_j^{(m+1)}(z_\ell) = \gamma_{j\ell}, \quad \ell = 1,\dots,m_j, \quad j = o,\dots,n-1. \tag{4.46}$$

REFERENCES

[1] Adams, R. A., Sobolev Spaces, Academic Press, New York, 1975.

[2] Aubin, J. P., Approximation of Elliptic Boundary-Value Problems, Wiley Interscience, New York, 1972.

[3] Babuška, I. and Aziz, A. K., Survey Lectures on the mathematical foundations of the finite element method, in "The Mathematical Foundations of the Finite Element Method with Applications to Partial Differential Equations", ed. by K. Aziz, Academic Press, 1972, p. 3-359.

[4] Begehr, H. and Gilbert, R. P., Über das Randwert-Normproblem für ein nichtlineares elliptisches System, Proc. Conf. Function Theoretic Methods for Partial Diff. Equs., Darmstadt, 1976, Lecture Notes in Mathematics 561, Springer, Heidelberg (1976), 112-122.

[5] Begehr, H. and Gilbert, R. P., Randwertaufgaben ganzzahliger Charakteristik für verallgemeinerte hyperanalytische Funktionen, Applic. Analysis 6 (1977), 189-2o5.

[6] Bers, L., John, F. and Schechter, M., Partial Differential Equations, Interscience Publ., New York, London, Sidney, 1964.

[7] Bohl, E., Stabilitätsungleichungen für diskrete Analoga nichtlinearer Randwertaufgaben, in: Numerische Behandlung von Differentialgleichungen, ed. Ansorge, Collatz, Hämmerlin, Törnig, ISNM 27 (1975), Birkhäuser, Basel und Stuttgart, p. 9-28.

[8] Bojarski, B., Theory of generalized analytic vectors, Ann. Polon. Math. 17 (1966), 281-320, (Russian).

[9] Bramble, J. and Schatz, A., Rayleigh-Ritz-Galerkin-Methods for Dirichlet's problem using subspaces without boundary conditions, Comm. Pure Appl. Math. 23 (1970), 653-675.

[10] Douglis, A., A function-theoretic approach to elliptic systems of equations in two variables, Comm. Pure Appl. Math., VI (1953), 259-289.

[11] Fix, G. J. and Strang, G., An Analysis of the Finite Element Method, Prentice Hall, Englewood Cliffs, New York, 1973.

[12] Gilbert, R. P., Pseudohyperanalytic function theory, Berichte Ges. Math. u. Datenverarbeitung mbH., Bonn 77 (1973), 53-63.

[13] Gilbert, R. P., Nonlinear boundary value problems for elliptic systems in the plane, Proc. Internat. Conf. Nonlinear Systems Appl. 1976, Techn. Report G3, University of Delaware, 1976.

[14] Gilbert, R. P. and Hile, G. N., Generalized hypercomplex function theory, Transc. Amer. Math. Soc. 195 (1974), 1-29.

[15] Haack, W. and Wendland, W., Lectures on Pfaffian and Partial Differential Equations, Pergamon Press, 1972.

[16] Höhn, W., Numerische Behandlung nichtlinearer Neumann-Probleme mit finiten Elementen, Dissertation TH Darmstadt, 1976.

[17] Nirenberg, L., An application of generalized degree to a class of nonlinear problems, Coll. Analyse Fonctionelle, Liège, 1972, pp. 57-74.

[18] Nitsche, J. A., Zur Konvergenz von Näherungsverfahren bezüglich verschiedener Normen, Numer. Math. 15 (1970), 224-228.

[19] Nitsche, J. A., On L_∞-convergence of finite element approximations to the solution of a nonlinear boundary value problem, in: Proceedings, Conference on Numerical Analysis 1876, Dublin, Ireland, to appear.

[20] Schäfer, H., Über die Methode der a-priori-Schranken, Math. Annalen 129 (1955), 145-416.

[21] Tutschke, W., Lösung nichtlinearer partieller Differentialgleichungssysteme erster Ordnung in der Ebene durch Verwendung einer komplexen Normalform, Math. Nachr. 75 (1976), 283-298.

[22] Tutschke, W., The Riemann Hilbert problem for nonlinear systems of differential equations in the plane, to appear.

[23] Vekua, I. N., Generalized Analytic Functions, Pergamon, 1962.

[24] Wacker, H. J., Eine Lösungsmethode zur Behandlung nichtlinearer Randwertprobleme, in "Iterationsverfahren, Numerische Mathematik, Approximationstheorie", ISNM 15, Birkhäuser, 1970, 245-257.

[25] Walter, W., Differential and Integral Inequalities, Springer Verlag, Berlin, Heidelberg, New York, 1970.

[26] Wendland, W., An integral equation method for generalized analytic functions, Proceedings, in Springer Lecture Notes 430 (1974), 414-452.

[27] Wendland, W., Elliptic systems in the plane, Pitman Publishing Inc., London, 1978.

[28] Wendland, W., On a class of semilinear boundary value problems for certain elliptic systems in the plane, Academy of Science of the USSR, ed. N. Bogolubov, to appear (Preprint Nr. 295, 1976, Fachbereich Mathematik, TH Darmstadt).

[29] Wloka, J., Funktionalanalysis und Anwendungen, De Gruyter, Berlin, New York, 1971.

[3o] Rannacher, R., Punktweise Konvergenz der Methode der finiten Elemente beim Plattenproblem, manuscr. math. 19 (1976), 4o1-4o6 (added in proof).

A B C 8 D 9 E 0 F 1 G 2 H 3 I 4 J 5